The North American Galloway Cattle Herd Book
Containing Pedigrees of Purebred Galloway Cattle

by North American Galloway Cattle Association

with an introduction by Jackson Chambers

This work contains material that was originally published in 1883.

This publication is within the Public Domain.

This edition is reprinted for educational purposes
and in accordance with all applicable Federal Laws.

Introduction Copyright 2018 by Jackson Chambers

Self Reliance Books

Get more historic titles on animal and stock breeding, gardening and old fashioned skills by visiting us at:

http://selfreliancebooks.blogspot.com/

Introduction

I am pleased to present another title in the "Cattle" series.

The work is in the Public Domain and is re-printed here in accordance with Federal Laws.

As with all reprinted books of this age that are intended to perfectly reproduce the original edition, considerable pains and effort had to be undertaken to correct fading and sometimes outright damage to existing proofs of this title. At times, this task is quite monumental, requiring an almost total "rebuilding" of some pages from digital proofs of multiple copies. Despite this, imperfections still sometimes exist in the final proof and may detract from the visual appearance of the text.

I hope you enjoy reading this book as much as I enjoyed making it available to readers again.

Jackson Chambers

PREFACE.

The Agricultural and Arts Association have now the pleasure of presenting the first volume of the North American Galloway Herd Book to the public. Although the first importation of these useful and hardy animals was made in 1853 by the Messrs. Graham, of Vaughan, the registration of them was not commenced until May in 1872, and up to 1874 was confined entirely to Canada. At that date the Americans and Canadians who had moved to the United States began to take some over, and since then have taken nearly all that were in the country, so that the old name of "Ontario Galloway Herd Book" was found to be quite too local.

At a meeting of the Galloway Breeders in Chicago, on 24th of Nov., 1882, during the Fat Stock Show of 1882, the following resolutions were unanimously adopted :

1. There shall be an Association, comprising all the breeders of Galloway cattle in North America, for the purpose of promoting the interests of the breed.
2. That the principal meetings of such Association should be at such time and place as shall be most convenient for the majority of the breeders.
3. That such association should arrange for the publication of a record of the pedigrees of pure-bred Galloway cattle.
4. That such Association should arrange for the organization of Branch Societies in such State or Province where Galloway cattle are bred, and that such bodies should have representation in the meetings of the parent organization.
5. That as the majority of Galloway cattle now in America are recorded in the "Ontario Galloway Stock Register of Pure-bred Galloways," it is desirable to confer with the Agricultural and Arts Association of Ontario as to whether they are willing to allow this Association to supervise all pedigrees, and publish them in a book, to be called "The Galloway Herd Book of North America."
6. That the President and officers elected at the recent meeting in Kansas City be requested to act in accordance with the suggestions made in these resolutions.

President, F. McHardy, Emporia, Kansas, U. S.
Secretary, D. McCrae, Guelph, Ont.

The conditions requested at this meeting were consented to by the Council of the Agricultural and Arts Association at a future meeting, and this volume is now the result. The Editor has been ably assisted

PREFACE.

by Mr. D. McCrae, of Guelph, on behalf of the North American Breeders Association, who has prepared a very valuable history of the Galloway cattle, published in this volume, which will be appreciated, I am sure, by the breeders of this fine race of cattle.

Every animal published has a satisfactory pedigree with the exception of one, Jacob Benner's Bull [196]; and I have since ascertained from W. Torrance, of the Co. of York, that he thinks he had the bull and sold him to a gentleman by the name of Benner, and if so he would be from *Black Bess* [6]; he was no doubt a pure Galloway.

On page 168, Craignarget [611] is published as a cow when he really is a Scotch bull, *Craignarget* [611]; the purchasers had better mark this in their books to prevent mistakes in future.

In order to distinguish imported animals, they are printed in italics throughout the volume.

Blank forms, free of charge, for entry, giving full particulars, can be procured at any time by writing to the editor,

HENRY WADE,
Toronto, Ont.

HISTORY OF GALLOWAYS.

By Mr. D. McCrae, Guelph.

Galloway, which has given its name to a valuable breed of black polled cattle, is an ancient regality or lordship lying in the south-west of Scotland. The word is derived from *Gallovid*, which in old Scots signifies "a Gaul." The Gauls are said to have been the first inhabitants of this part of Scotland. They, under their "Reguli" or petty kings, enjoyed a sort of independence, had their own special laws and judges. For many centuries their national councils met on the "Mote of Urr," were presided over by their lords, and passed and promulgated such new laws as might be required. The last of the old line of rulers was Alan, Lord of Galloway, who was buried in Dundrennan Abbey in 1233. The title passed into the house of Douglas of Thrieve, and the whole district was finally annexed to the Crown of Scotland in 1455. Galloway at this time extended as far as the Firth of Forth, and in addition to its modern boundaries comprised part of Ayrshire, Lanark, Renfrew and Dumfriesshire. The name now embraces only the Stewartry of Kirkcudbright and the shire of Wigton. It is about ninety miles in length by forty in breadth. Its surface is very undulating. Near the sea and along the valleys of its numerous rivers are many fertile glens of good land, yielding rich pasturage and good crops. On the higher ground the surface is more irregular, with lochs, stretches of moorland, high granitic hills, mosses and heathery mountain ranges. Originally the whole of this land was covered with dense forests, principally oak. It was all thickly wooded in the days of the Romans when they marched through it, made their roads, raised their forts, and feasted on good Galloway beef. Through these forests roamed many wild cattle generally supposed to be the progenitors of the modern breed of Galloways. Youatt, writing in 1840 of the cattle of Great Britain, says: "There appears to be the remnants of two distinct breeds of aboriginal cattle, one in the parks of Chillingham, in Northumberland; the other at Chatelherault, in Lanarkshire. The first are middle-horned, and the second are polled. The continuation of the first we have in the Devon, the Hereford, the Sussex and the Highland cattle. The others would appear to survive in the Galloways, the Angus humlies, the Suffolk and the Norfolks." This seems to be a reasonable deduction, and we conclude that these modern breeds of polled cattle are the lineal descendants of those wild aboriginal cattle which roamed in prehistoric times in the forests of Britain. This wild polled herd in Lanarkshire was not the only one in Britain. During the early part of the century there were several herds of wild polled cattle in England. Yorkshire, Nottingham, Lancashire and Cheshire all had representatives of these polled cattle. Nor were the polls confined to Britain. Herodotus describes the domestic cattle of the Scythians as being hornless. In Austria a fine herd of red polled cattle has existed for very many years—just how long is not known; their origin is lost in antiquity. Polled cattle are common in Norway. Iceland has a breed hornless, of "small size, but very fat and good." Ireland had at one time a breed of polled cattle. In the north of Scotland, Dr. Johnson, in his trip in 1773, saw several humlies." Strathmore had in very early times hornless cattle which are represented on the sculptured stones of Meigle.

Galloways are all hornless or polled. Some writers say that formerly they were mixed, some being horned and others polled. The writer has made diligent search to verify this, and has been unable to find any proof that Galloways were ever horned. The universal testimony amongst breeders of Galloways is that a pure-bred Galloway never had any trace of horns or even scurs. Such old breeders as James Grierson, of Caigton; Thos. Biggar, of Chapelton; John Cunningham, of Whitecairn; the late Peter Kerr, senior, of Bellymack, and Alexander Halliday, of Culcagrie, gave their testimony very strongly on this point. They all say that no pure Galloway ever had horns. Oral testimony handed down to these men from the Galloway breeders of the last century is valuable and reliable. We therefore conclude that while there were horned cattle in Galloway, yet there were none of the native breed, the pure Galloway, with horns. Professor Low, of Edinburgh, author of a valuable work on the domestic animals of Great Britain, and who very carefully investigated this matter, says: "Some earlier notices rather conduct us to the conclusion that the absence of horns has been for a long period a distinctive characteristic of the race." So marked is this polled character, that the produce of a pure-bred Galloway bull with any breed of horned cows should give polled calves. No other breed of polled cattle will equal the Galloway in this respect.

The breeding of cattle has been from time immemorial a principal object of the Galloway farmers. A compiled history of Scotland alluding to the time prior to and including the reign of Alexander III. (1249), says: "Black cattle were also reared in great numbers during the Scoto-Saxon period. The dairy was a considerable object of attention in the early ages of Scottish history; and cheese had been made in great abundance. As the people lived much on animal food, the cattle were all consumed within the land, while their skins formed a considerable article of export." George Buchanan, tutor to James I., of England, writing about 1566, says of Galloway: "It is more fruitful in cattle than in corn." Hector Boece (1570), writing of Galloway, says: "In this region ar mony fair ky and oxin of quhilk the flesh is right delicius and tender." Ortelius, the historian, writing in 1573, says: In Carrick (then part of Galloway) are oxen of large size, whose flesh is tender, sweet and juicy." This brings us to consider the excellence of the Galloway beef, which we see was acknowledged and recorded many hundred years ago. Galloways are pre-eminently a beef-producing breed. Their flesh is mottled or marbled fat and lean intermixed, and it was this quality which gave them their ancient fame, and which led to their being bred specially to supply the markets of England with beef of extra quality. After the union of England and Scotland this trade developed into large proportions. Many of the herds in those days were of large size. The author of "A Journey Through Scotland," London, 1723, mentions the flocks and herds of "Alexander Mackie, of Palgowan, who keeps at least ten thousand sheep on these mountains, besides an incredible number of black cattle. Not far from this is the famous mountain of Cairnsmure, full of deer and wild cattle. To the southwest stands a handsome seat, called the Caily, belonging to Alexander Murray, of Brochton, with a large park which feeds one thousand bullocks, that he sends once every year to the markets of England." Most of the Galloway bullocks were sent into Norfolk and Suffolk to the autumn fairs at St. Faith's and Hampton, there to be sold to the English graziers, who fed them for Smithfield and other English markets. They were sent on foot, grazing as they went, and usually took about three weeks on the road. The object of the breeder was to raise a good uniform lot, and as many as he could winter. Calves were never killed, and were usually allowed to suck the dam. On rough hill pastures they would receive most of the milk, but on the lower richer pastures the dairymaid often took half, and the calves did well on this limited share. Young cattle were seldom housed, but ran on the hills all winter. They were on the pastures day and night, and in times of deep snow received feed of coarse hay, sent to the fields on a horse's back. At three years old they were ready to go south. If a few of the two year olds were good enough to go with the drove, so much the better; they went. A good uniform lot was wanted, excellence in one or two individuals was not desired; this did not conduce to marked improvements being made in the breed. An extra good animal was far more likely to go south a two year old than be kept for the improvement of the breed. These droves, when finished on the richer pastures of England, brought the top price on the Smithfield market. Robert Brown, Esq., original editor of the *Farmer's Magazine*, says: "The Galloway cattle sell higher in Smithfield market per stone than any other breed, owing to the fat being laid upon the most valuable parts, which seems to be a quality of the greatest excellence when the value of feeding cattle is to be ascertained. The hides of this breed are not so thick as those of the long-horn, nor so thin as those of the short-horn breed; but their beef is well marbled or mixed with fat, and therefore, in point of quality, much superior to that of either. Hence the demand for Galloway cattle is always very considerable, and it is no uncommon thing to see one of these little bullocks sell for as much money in Smithfield as a Lincolnshire ox of double its weight. The cows of this breed are good milkers, and the milk which they yield is of a very rich quality. Several unsuccessful attempts have been made to amend this breed by crossing with bulls from other counties, but the result has satisfactorily shown that the real original polled breed can only be preserved in perfection by selecting bulls at home of the best figure and properties." There is no breed of cattle which can lay such claim to the title "pure-bred" as the Galloways. It has never been said in any well-informed quarter that the Galloway is not an original and distinct breed of cattle. It has had no mixture with other breeds. All the improvements in the breed have come from within by careful selection. There is no breed of cattle which can more truly be said to be indigenous to the country and incapable of improvement by any foreign cross. Youatt says: "The short-horns, almost everywhere else, have improved the cattle of the districts to which they travelled, but even in the first cross the short-horns have done little good in Galloway, and as a permanent mixture the choicest southern bulls have manifestly failed. The intelligent Galloway breeder is now perfectly satisfied that his stock can only be improved by adherence to the pure breed." And this experience has been arrived at by long and careful experiments. Many of the very best breeders have tried experiments with the most likely breeds. The Long-Horns, old Westmoreland breeds, the Short-Horns, Devons, West Highlands, polled Angus—all have been tried, and the results have all been disappointing. Mr. Cranston, manager for the Duke of Buccleuch, is at present working on a cross with the West Highland cattle, but none of these have been a success. The attempt to improve by a cross with polled Angus blood was as great a failure as any. The first cross is usually a very superior animal, and in many cases feeds better and matures quicker than either of the parents. But they require to be kept and bred for the butcher only, as almost invariably the next cross is not so good. At the present day these crosses are much in favor in England. Amongst the numerous dairy farmers in Ayrshire and Wigtonshire many Galloway bulls are kept where the calves are to be fed for beef, and the Ayrshire Galloway cross are found to feed well. One of the very best feeding animals is the first cross between the Galloway and Short-Horn. These are all polled, but not usually all black; many of them have a mixture of white, and are popularly known as blue-greys. They are much esteemed by the English feeders and butchers. In these crosses the wonderful power of the Galloways is clearly shown. The half of the cross-bred animals will have a strong resemblance to the Galloway, and some will be so very much like the pure Galloway as to deceive any but an expert. And in this very potency and power lies a great danger. Many animals only half-bred may be sold by unscrupulous persons as pure-bred, and, as a consequence, the results will be disappointing. In the next generation the baser blood will come out, and naturally the breed will be condemned when it has never had a fair trial.

While for many years black has been the predominant color, it was not the only one amongst the Galloways. In the quotations already given, black is several times spoken of as the color of the cattle of Galloway. The report of the Scottish Agricultural Society (1743) says: "The Earl of Stair was breeding black cattle on the Galloway hills better than any raised in either the Lothians or Fyfe." Breeders have thought that the blacks were hardier and stood the winters better, and so selected them usually for their breeding stock. So closely have they kept to that color that it is now rare to find a Galloway of any other. But others may be just as well and purely bred as the blacks. There is no diversity of opinion on this point. It is acknowledged by all old breeders that last century many herds were mixed in color. Any appearance of horn or scur will be condemned at once as clear evidence of impure blood, but a red, a black and white, or a dun will be acknowledged as quite pure. This point comes clearly out in crossing with other breeds. If the cross is made with an old established breed the probability is that quite half the calves will show variety of colors, while very probably not any will show the slightest symptom of horns. And the Galloway black is not a pure black. The calves when first dropped should be a dark mahogany brown; the under coat of hair should always have this brown tinge, which is more clearly seen when it is being shed. A very deep black is oftener found in the half-breeds, which are frequently quite black. The brown tinge is one of the best signs of purity of blood. Many families have white on the belly, usually a small patch on the udder. Others have a few white hairs scattered through the coat, not usually noticed, and it is wonderful how these run in families for generations. Belted ones have a band of white around the middle; they are still met with but not nearly as common as they were fifty years ago. This belted race had a good reputation as milkers. Another pure herd had white faces, and others had a white mark along the back very much after the style of the modern Herefords. Brindled ones were formerly common, and drab or dun, red, and red and white, were also met with. This being the case, by careful selection any breeder may work back to a breed of pure Galloways of different colors, provided, of course, that he can get those away from the pure black to start with.

Galloways are docile and quiet. Though very active and hardy, they are easily managed, though able to hold their own with other breeds. They are very courageous, and if annoyed by dogs or wild animals will act in concert, form up in a crescent, and jointly attack in that formation, heads to the centre and the flanks gradually coming around. One or two Galloways in a field with sheep prevent any danger from dogs worrying them. They will attack the dogs at once.

Regarding the points of Galloway cattle we have that published by Aiton in 1811. He says: "The general aspect of Galloway breed is robust, fierce, lively and spirited. The muscles and tendons are large and strong, which enables them to travel with ease and endure much fatigue. They should have—

HEAD—Short, square, narrow at muzzle and without horns.
    Eye—Quick and lively.
NECK—Rather bended upwards (bulls), small near the head, loose skin under the chops a blemish.
BODY—Straight and round, plenty of room for the lungs.
    Back—Straight from ear to rump, broad and level over the kidneys, and no rise at the root of the tail.
    Ribs—Long.
    Chest—Round.
    Belly—Moderate size and pelvis large.
    Flank—Full, broad at the head of the loins over the fundament.
    Hams—Full and fleshy.
    Udder—Small.
    Teats—Short.
LEGS—Short and straighter at the hind knee-joint than those of other breeds.
TAIL—Thick, short and bushy with hair.
HIDE—Thick.
HAIR—Long and shaggy—formerly curled hair was preferred, but it has been discovered that cows whose hair is curled are slow feeders.

They are of tolerable size and very handsome, are spirited, strong, very healthy and hardy, and no cattle whatever feed better or yield beef that is more relished at table."

To show the changes that have taken place since, here is a statement of the characteristics of a typical animal of the Galloway breed, drawn up by the Council of the Galloway Cattle Society of Great Britain, 18th April, 1883:

COLOR—Black, with a brownish tinge.
HEAD—Short and wide, with broad forehead and wide nostrils; without the slightest symptoms of horns or scurs.
    Eye—Large and prominent.
    Ear—Moderate in length and broad, pointing forwards and upwards with fringe of long hairs.
NECK—Moderate in length, clean, and filling well into the shoulders; the top in a line with the back in a female, and in a male naturally rising with age.

BODY—Deep rounded, and symmetrical.
- *Shoulders*—Fine and straight, moderately wide above; coarse shoulder points and sharp or high shoulders are objectionable.
- *Breast*—Full and deep.
- *Back and Rump*—Straight.
- *Ribs*—Deep and well sprung.
- *Loin and Sirloin*—Well filled.
- *Hook Bones*—Not prominent.
- *Hind Quarters*—Long, moderately wide, and well filled.
- *Flank*—Deep and full.

THIGHS—Broad, straight, and well let down to hock; rounded buttocks are very objectionable.
- *Legs*—Short and clean, with fine bone.
- *Tail*—Well set on, and moderately thick.

SKIN—Mellow, and moderately thick.
- *Hair*—Soft and wavy, with mossy undercoat; wiry or curly hair is very objectionable.

The differences between these statements are not very great. In the modern scale prominence is given to the ear. Narrow muzzles are not now looked upon with favor. Hams may still be full and fleshy, but must be straight and square; rounded ones are objectionable. The thick skin has now given place to "moderately thick," and softness of hair with mossy undercoat is now a special feature. Some animals are without this thick mossy undercoat, which should, in the very best hides, have an appearance and feeling much akin to a sealskin jacket. The advantage of such a covering will be seen by all breeders. In cold or windy weather it has a great warmth, and in wet days will throw off almost any amount of rain. And now for the points which should be avoided, and which may be of as much benefit to the beginner in Galloway breeding as the former scales of desirable points. For this list I am indebted to Mr. James Biggar, "The Grange," Dalbeattie, Scotland. Galloways should not have

1. Long narrow head with high crown.
2. Narrow tapering muzzle.
3. Long drooping ears.
4. Small deep-set eyes.
5. Small light neck.
6. Light scraggy breast.

7. High narrow shoulders.
8. Flatness behind shoulders.
9. Light fore or back ribs.
10. Square and prominent hook bones.
11. High or drooping rumps.
12. Weak or slack loins.
13. Rounded buttocks.
14. Fleshy double thighs.
15. Big coarse bones.
16. Thick stiff skin.
17. Hard wiry or too curly hair.
18. Black hard hair without soft undercoat.

These defects should be avoided as much as possible by the careful breeder. It is true it will be difficult to get a herd without some of these faults, but the knowledge of what should be shunned will assist in bringing a herd up to a good place in the standard of excellence. Very large animals are not desirable—the breed is not a large one. Galloways cannot usually be raised to such heavy weights as some other beef breeds, and the largest are not usually the best or most desirable. A good medium-size, quite attainable, is quite as desirable for the butcher. Short-legged animals, with small fine bone, are often the best. These will look very small in the field when with their longer-legged neighbors, but when they come to the scales the difference is often in their favour. Nor can they be made to put on as much outside fat as some other breeds. While this is a quality going a great way in a show yard, it is not one that commends itself to the consumer. The Galloway puts the fat on the most valuable parts, and the meat when cut is found to have the fat intermixed with the lean, a most valuable and desirable quality. There is no beef in the world can beat that of a well fed Galloway. Mr. William McCombie, the great breeder of polled Aberdeen and Angus cattle, in his work on cattle and cattle feeders, says: "There is no other breed worth more by the pound weight than a first-class Galloway." An eminent English butcher, Mr. Joseph Hall, writing in 1883, says: "I think there is no cattle equal to the polled Galloway as a beef-producing breed. The quality of beef is superior to any I ever killed, and the offal more profitable. As to early maturity I have killed polled Galloway heifers fifteen months old, costing £21 ($105) each. No class of cattle make such heavy weights with so little bone, and the lean meat cuts so thick and fine in the grain that it pleases both butcher and consumer. I have been in the trade nearly fifty years, and found Galloway cattle second to none." This shows that for early maturity they are equal to the best specimens of any breed. At present in Britain the best feeders send their animals to the market at from two years to two years and ten months old. They can be fed to a good paying weight at an early age, a very desirable quality. There is little doubt that Galloway beef either as half or pure bred, will in a few years be quoted at as much more than usual price of ordinary beef in the markets of Kansas City, Chicago and New York as it does to-day in those of Liverpool, Manchester and London. Then, as to transport, they can stand long journeys by road or rail better than other breeds. This hardy quality, arising from their great muscular development, is quite as useful on the western plains where they may have to travel considerable distances to gather their food, to or from water, or on their way to market. On the North-West ranches they are found more active and hardy than other breeds. In cold wintry weather they will be first out scraping off the snow to get at the grasses below. Their good coat of hair gives them great advantages in cold or wet climates, and their moderately thick hides and abundant long hair protect them in warm weather from the attacks of insects. They acclimatize in the south and west very readily, feed on a larger number of plants than most other thorough-breds, and their courage and hardihood are found to help them greatly amongst large herds. While they are not a milking breed, and their milk is usually limited in quantity, it is exceedingly rich in quality. Galloway cows can always raise sleek fat calves, and make good nurses, looking well after their progeny.

Though the Galloways are the oldest of the pure-bred breeds of British cattle, their herd book records are very modern. Unfortunately all the papers and documents which had been collected relating to the breed, as well as the pedigrees, were lost by the fire which destroyed the Highland and Agricultural Society's Museum and Records in Edinburgh, in 1851. Since that time a new book was begun, and the first volume of the Polled Herd Book was published in 1862. This was compiled and arranged by Mr. Edward Ravenscroft, and contained pedigrees of animals of the polled Aberdeen, Angus and Galloway breeds. Few of the breeders of Galloways joined in it. In 1878 the Galloway Cattle Society published its first volume of pedigrees, arranged by the able Secretary, Rev. John Gillespie, of Monswald Manse, Dumfries, No pains were spared to make this record full and complete. Several volumes have followed since that time. The one for this year will be much larger than any of its predecessors. The Scottish Society have a committee of old and experienced breeders before whom all pedigrees come before being finally recorded. It is to be hoped that the Galloway breeders of this continent will take similar pains to have all records pure, and with careful earnest work the Galloways will soon have the place they deserve—in the front rank of the beef breeds of North America.

NORTH AMERICAN GALLOWAY HERD BOOK.

VOL. I.

BULLS.

[10] *JOCK.*

Calved in 1852.

The property of Mr. Graham, Vaughan, Co. York, Ont.; imported in 1853 from the herd of Mr. Rogerson, Leighthan Hall, Dumfries, Scotland.

[15] DRED.

Calved March 24th, 1861.

Bred by Mr. John Snell, Chinguacousy, Co. Peel, Ont.; the property of Mr. Thomas McCrae, Guelph, Ont.

Sire *Moss Trooper* [20].
Dam *Sall* [14] Imported in 1856.

[17] *BLACK JOCK.*

Owned by Mr Carruthers, Kirk Hill, Dumfries, Scotland.

[18] BLACK JOCK.

Calved April 16, 1854.

Bred by Mr. John Fleming, Vaughan, Co. York, Ont.; the property of Mr. John Snell, Chinguacousy, Co. Peel, Ont.

Sire *Black Jock* [17].
Dam *White Bag* [5] Imported in 1853.

[19] MARQUIS.

Calved April, 1857.

Imported in dam, and the property of Mr. James Graham, Vaughan, Co. York, Ont.

Sire *The Pilgrim* [32] (32).

Dam *Topsey* [13] Imported in 1856.

[20] MOSS TROOPER.

Calved April, 1857.

Imported in dam, by Mr. W. R. Graham, Vaughan, Co. York, Ont.; the property of Mr. Jas. Graham, of the same place.

Sire *Scotland* [21].

Dam *Beauty* [11] Imported in 1856.

[21] SCOTLAND.

Owned by Mr. Carruthers, Kirk Hill, Dumfries, Scotland.

[23] LORD BYRON.

Calved October, 1862.

Bred by and the property of Mr. James Graham, Vaughan, Co. York, Ont.

Sire Black Jock [18].

Dam *Heather Bell* [12] Imported in 1856.

[24] TAM O'SHANTER.

Calved June, 1861.

Bred by Mr. James Graham, Vaughan, Co. York, Ont., and the property of Mr. George Robson, Markham, Co. York, Ont.

Sire Black Jock [18].

Dam *Heather Bell* [12] Imported.

[26] McQUHOON.

Calved July, 1862.

Bred by Mr. James Graham, Vaughan, Co. York, Ont., and the property of Mr. George Roddick, Cobourg, Ont.

Sire Black Jock [18].

Dam *Chloe* [4] Imported.

BULLS.

[28]   WILLIAM WALLACE, *ALIAS* CLOVER HILL.

Calved May, 1876.

Bred by Mr. John Fleming, Vaughan, Co. York, Ont., and the property of Mr. John McLean, Clover Hill, Co. Simcoe, Ont.

Sire Black Jock [18].
Dam *White Bag* [5] Imported.

[29]   GRAMPIAN.

Calved May, 1857.

Bred by and the property of Mr. John Fleming, Vaughan, Co. York, Ont.

Sire Black Jock [18].
Dam *White Bag* [5] Imported.

[32] (32)   *THE PILGRIM.*

The property of Mr. James Grierson, Caigton Hall, Dalbeattie, Scotland.

Sire *Geordie of Riggfoot* (234).

[37]   BLACK TOM.

Calved in 1855.

Bred by and the property of Mr. W. R. Graham, Vaughan, Co. York, Ont.

Sire Black Jock [18].
Dam *Bonny* [1] Imported.

[40]   YOUNG MATCHLESS.

Calved June, 1862.

Bred by and the property of Mr. Arthur McNeil, Vaughan, Co. York, Ont.

Sire Black Tom [37].
Dam Young Bonny [39] by Black Tom [37].
2nd d. *Bell* [9] Imported.

[41]   YOUNG BLUCHER.

Calved February 1st, 1863.

Bred by and the property of Mr. Arthur McNeil, Vaughan, Co. York, Ont.

Sire Black Jock [18].
Dam *Bell* [9] Imported.

[44] CURLY BOB.

Calved May, 1861.

Bred by and the property of Mr. John Torrence, Vaughan, Co. York, Ont.

Sire Black Jock [18].
Dam *Venus* [3] Imported.

[45] JOCK 2ND.

Calved April 30, 1857.

Bred by Mr. W. R. Graham, Vaughan, Co. York, Ont.; the property of Mr. Edward Welding, same place.

Sire *Jock* [10].
Dam *Jet* [2] Imported.

[46] FRED DOUGLAS.

Calved March 24, 1860.

Bred by Mr. John Snell, Chinguacousy, Co. Peel, Ont.; the property of Mr. George Anderson, Stanley, Ont.

Sire Black Jock [18].
Dam Bonny [27] by *Jock* [10].
2nd d. *White Bag* [5] Imported.

[47] NELSON.

Calved November, 1860.

Bred by and the property of Mr. George Cotter, Nelson, Co. Halton, Ont.

Sire Black Jock [18].
Dam *Venus* [3] Imported.

[48] (151) *MOSS TROOPER 2nd.*

Bred by and the property of Mr. Joseph Beattie, Newbie House, Annan, Scotland.

Sire *Moss Trooper* (296).
Dam *Maggie of Dalbeattie* (767) by *Fergy* (19.)

BULLS.

[57]   PRINCE OF WALES.

Calved October, 1859.

Bred by Mr. John Torrence, Vaughan, Co. York, Ont.; the property of Col. E. W. Thompson, Aikenshaw, Co. York, Ont.

Sire Black Jock [18].
Dam *Beauty* [11] Imported.

[58]   UNCLE TOM.

Calved April 8th, 1861.

Bred by Mr. John Snell, Chinguacousy, Co. Peel, Ont.; the property of Mr. Thos. Armitage, Colchester.

Sire *Marquis* [19].
Dam Bonny 2nd [36] by Black Jock 18.
2nd d. *Bonny* [1] Imported.

[62]   MALCOLM

Calved January 1st, 1862.

Bred by and the property of Mr. John Snell, Chinguacousy, Co. Peel, Ont.

Sire Fred Douglas [46].
Dam Empress [50] by *Jock* [10].
2nd d. *Beauty* [11] Imported in 1856.

[63]   *YOUNG SCOTLAND.*

Calved January 15th, 1861.

Bred by and the property of Mr. John Snell, Chinguacousy, Co. Peel, Ont.

Sired in Scotland.
Dam *Blooming Heather* [53] by *Moss Trooper 2nd* [48] (151).
2nd d. *Mary* by *Fergy* (19).
3rd d. *Jane* by *Bob*. All in Scotland.

[64]   DUNCAN

Calved March 15th, 1862.

Bred by and the property of Mr. John Snell, Chinguacousy, Co. Peel, Ont.

Sire Fred Douglas [46].
Dam *Sall* [14] Imported in 1856.

[66] ROBERT BRUCE.

Calved December 10th, 1862.

Bred by and the property of Mr. John Snell, Chinguacousy, Co. Peel, Ont.

Sire Dred [15].
Dam Bonny [27] by Black Jock, 18.
2nd d. *White Bag* [5] Imported in 1853.

[67] WILLIAM WALLACE.

Calved January 14th, 1864.

Bred by and the property of Mr. John Snell, Chinguacousy, Co. Peel, Ont.

Sire Fred Douglas [46].
Dam Empress [50] by *Jock* [10].
2nd d. *Beauty* [11] Imported in 1856.

[68] RODERICK DHU.

Calved January 31st, 1863.

Bred by and the property of Mr. John Snell, Chinguacousy, Co. Peel, Ont.

Sire Fred Douglas [46].
Dam Jessie [49] by Black Jock [18].
2nd d. *Maggie* [48] by *Moss Trooper 2nd* [48] (151).

[69] *THE EARL.*

Bred and owned by the Earl of Selkirk, St. Mary's Isle, Kircudbrightshire, Scotland.

Dam *Beauty* [151] by *Peter* [116].

[70] GALLOWAY TAM.

Calved Spring of 1857.

Bred by and the property of Mr. George Roddick, Cobourg, Co. Northumberland, Ont.

Sire *Young Moss Trooper* [73].
Dam *Beauty* [151] Imported.

BULLS.

[71] SAMBO.

Calved February 20, 1863.

Bred by and the property of Mr. John Dunlop, East Zorra, Co. Oxford, Ont.

Sire Morrach [78].

Dam Heather Bell 2nd [25] by Black Tom [37].

2nd d. *Heather Bell* [12] Imported in 1856.

[72] NORVAL.

Calved in 1854.

Bred by Mr. W. R. Graham, Vaughan, Co. York, Ont.; the property of Mr. Allan Wilcox, Co. Peel, Ont.

Sire *Jock* [10].

Dam *Bonny* [1] Imported in 1853.

[73] *YOUNG MOSS TROOPER*

Bred by and the property of Mr. Beattie, Newbie House, Annan, Scotland.

Sire *Moss Trooper* (296).

[74] (296) *MOSS TROOPER.*

Calved in 1846.

Bred by Mr. George Graham, Riggfoot, Co. Cumberland; the property of Mr. James Beattie, Newbie House, Annan, Scotland.

Sire *Cumberland Willie* (160).

Dam *Maid of Lochfouin* (62) in Scotland.

[76] (234) *GEORDIE OF RIGGFOOT.*

Calved in 1847.

Bred by Mr. George Graham, Riggfoot, Cumberland; the property of Mr. W. & I. Sherman, Balig, Kircudbright.

Sire *Cumberland Willie* (160).

[78]   MORRACH.

Calved in 1858.

Bred by Mr. George Roddick, Cobourg, Co. Northumberland, Ont.; the property of Mr. John Dunlop, East Zorra, Co. Oxford, Ont.

Sire *Young Moss Trooper* [73].

Dam *Dandy* [69] by *The Earl* [69].

2nd d. *Black Jess* by *Geordie of Riggfoot* [76] (234).

3rd d. *Blackie* by *Galloway Tam*, (100).

[90]   CARIBOO.

Calved March 25th, 1865.

Bred by and the property of Mr. John Snell, Chinguacousy, Co. Peel, Ont.

Sire *Dryfe* [222].

Dam Susan [30] by Black Jock [18].

2nd d. *White Bag* [5] Imported in 1853.

[91]   EL HAKIM.

Calved March 4th, 1866.

Bred by and the property of Mr. John Snell, Chinguacousy, Co. Peel, Ont.

Sire *Victor* [122].

Dam Susan [30] by Black Jock [18].

2nd d. *White Bag* [5] Imported in 1853.

[92]   VICTOR HUGO.

Calved April 12th, 1866.

Bred by and the property of Mr. John Snell, Chinguacousy, Co. Peel, Ont.

Sire Victor [122].

Dam Flora, [22] by Black Jock [18].

2nd d. Shaw, [21] by *The Pilgrim* [32] (32).

3rd d. *Sall*, 14, Imported in 1856.

BULLS.

[93]   TONAWANDA.

Calved November 10th, 1866.

Bred by Mr. John Snell, Chinguacousy, Co. Peel, Ont.; the property of Mr. Thomas McCrae, Guelph, Co. Wellington, Ont.
Sire Black Jock [18].
Dam Lavina [55] by Black Jock [18].
2nd d. Empress [50] by *Jock*, [10].
3rd d. *Beauty* [11] Imported in 1856.

[94]   SAGINAW.

Calved November 25th, 1866.

Bred by Mr. John Snell, Chinguacousy, Co. Peel, Ont.; the property of Mr. Thomas McCrae, Guelph, Co. Wellington, Ont.
Sire Black Jock [18].
Dam Annie Lawrie [82] by Black Jock [18].
2nd d. Empress [50] by *Jock* [10].
3rd d. *Beauty* [11] Imported in 1856.

[98]   HONEST TOM.

Calved June 18th, 1862.

Bred by and the property of Mr. A. McNeil, Vaughan, Co. York, Ont.
Sire Black Tom [37].
Dam *Bell* [9] Imported in 1853.

[100] (100)   *GALLOWAY TAM.*

Bred by and the property of Mr. John Muir, Highbae, Castle Douglas, Scotland.

[101]   SELKIRK.

Calved February 20th, 1866.

Bred by and the property of Mr. Thos. McCrae, Guelph, Co. Wellington, Ont.
Sire Donald [123].
Dam Bonny [27] by Black Jock [18].
2nd d. *White Bag* [5] Imported in 1853.

[102]   JOHN A.

Calved May 12th, 1867.

Bred by and the property of Col. R. L. Denison, Dover Court, Toronto.
Sire Wonderful Lad [214].
Dam Salina [104] by Prince of Wales [57].
2nd d. *Helen* [103] Imported.

[106]   OUR JOHN.

Calved October 11th, 1865.

Bred by Mr. John Snell, Chinguacousy, Co. Peel, Ont.; the property of Mr. Wm. Hood, Guelph, Co. Wellington, Ont.
Sire Black Jock [18.]
Dam *Sall* [14] Imported in 1856.

[107]   UNCLE TOM.

Calved January 3rd, 1867.

Bred by and the property of Mr. Wm. Hood, Guelph, Co. Wellington, Ontario.
Sire *Pride of the Speed* [159].
Dam *Sall* [14] Imported in 1856.

[108]   GEORDIE.

Calved January 28th, 1868.

Bred by and the property of Mr. Wm. Hood, Guelph, Co. Wellington, Ont.
Sire *Pride of the Speed* [159].
Dam Nelly Grey [109] by Black Jock [18].
2nd d. *Maggie* [48] by *Moss Trooper 2nd* [48] (151).

[113]   UNCLE TOM.

Calved March 28th, 1858.

Bred by and the property of Mr. Wm. Roddick, Cobourg, Co. Northumberland, Ont.
Sire *Young Moss Trooper* [73].
Dam *Barbara* [113] Imported.

BULLS.

[115] CHUB.

Calved March 20th, 1867.

Bred by and the property of Mr. Arthur McNeil, Vaughan, Co. York, Ont.

Sire *Dryfe* [222].
Dam *Bell* [9] Imported in 1853.

[116] *PETER.*

Bred by and the property of Mr. William Sproat, Borness Bogue, Kircudbright, Scotland.

[117] TECUMSEH.

Calved October 21st, 1864.

Bred by and the property of Mr. John Snell, Chinguacousy, Co. Peel, Ont.

Sire Black Jock [18].
Dam Bonny [27] by *Jock* [10].
2nd d. *White Bag* [5] Imported in 1853.

[118] EL DORADO.

Calved November 25th, 1864.

Bred by and the property of Mr. John Snell, Chinguacousy, Co. Peel, Ont.

Sire Black Jock [18].
Dam Dairymaid [51] by Black Jock [18].
2nd d. Galloway Lass [42] by *Jock* [10].
3rd d. *Jet* [2] Imported in 1853.

[120] ST. CLAIR.

Calved November 21st, 1863.

Bred by and the property of Mr. John Snell, Chinguacousy, Co. Peel, Ont.

Sire Black Jock [18].
Dam Bonny [27] by *Jock* [10].
2nd d. *White Bag* [5] Imported in 1853.

[121]   MONITOR.

Calved December 10th, 1863.

Bred by and the property of Mr. John Snell, Chinguacousy, Co. Peel, Ont.

Sire Black Jock [18].
Dam Dairymaid [51] by Black Jock [18].
2nd d. Galloway Lass [42] by *Jock* [10].
3rd d. *Jet* [2] Imported in 1853.

[122]   VICTOR.

Calved February 6th, 1864.

Bred by Mr. John Snell, Chinguacousy, Co. Peel, Ont.; the property of the Michigan State Agricultural College, Lansing, Michigan, U. S.

Sire Black Jock [18].
Dam Blooming Beauty [54] by Black Jock [18].
2nd d. Galloway Lass [42] by *Jock* [10].
3rd d. *Jet* [2] Imported in 1853.

[123]   DONALD.

Calved March 13th, 1864.

Bred by Mr. John Snell, Chinguacousy, Co. Peel, Ont.; the property of Mr. Thomas McCrae, Guelph, Co. Wellington, Ont.

Sire Black Jock [18].
Dam Lavina [55] by Black Jock [18].
2nd d. Empress [50] by *Jock* [10].
3rd d. *Beauty* [11] Imported in 1856.

[127]   YOUNG DRYFE.

Calved June, 1864.

Bred by and the property of Mr. Alex. Mounsey, Etobicoke, Co. York, Ont.

Sire *Dryfe* [222].
Dam Susan [30] by Black Jock [18].
2nd d. *White Bag* [5] Imported in 1853.

BULLS.

[130] GARABALDI.

Calved May 5th, 1865.

Bred by and the property of Mr. Alex. Mounsey, Etobicoke, Co. York, Ont.

Sire Dred [15].
Dam Agnes [124] by Norval [72].
2nd d. Bess [125] by Marquis [19].
3rd d. *Chloe* [4] Imported in 1853.

[133] AIKENDRUM.

Calved in 1868.

Bred by and the property of Mr. John Wilson, Westminster, Co. Middlesex, Ont.

Sire Caraboo [90].
Dam Louisa [86] by Dred [15].
2nd d Flora [22] by Black Jock [18].
3rd d. Shaw [21] by *The Pilgrim* [32].
4th d. *Sall* [14] Imported in 1856.

[134] SALTFLEET.

Calved June 9th, 1859.

Bred by Mr. Joseph Jardine, Saltfleet, Co. Wentworth, Ont.; the property of Mr. William Walker, Saltfleet, Co. Wentworth, Ont.

Sire *Moss Trooper* [135].
Dam *Bonny* [136] Imported.

[135] *MOSS TROOPER.*

Bred by Mr. Carruthers, Kirk Hill, Dumfries, Scotland.

[136] BONNIE DUNDEE.

Calved February 8th, 1867.

Bred by and the property of Mr. Thomas McCrae, Guelph, Co. Wellington, Ont.

Sire Dred [15].
Dam Bonny [27] by *Jock* [10].
2nd d. *White Bag* [5] Imported in 1853.

[137] SIR JOHN A.

Calved March 17th, 1867.

Bred by and the property of Mr. Thomas McCrae, Guelph, Co. Wellington, Ont.

Sire Dred [15].
Dam Queen of Beauty [95] by Black Jock [18].
2nd d. *Black Bess* [6] Imported in 1853.

[138] SIR WILLIAM WALLACE.

Calved May 1st, 1868.

Bred by and the property of Mr. Thomas McCrae, Guelph, Co. Wellington, Ont.

Sire Dred [15].
Dam Bonny [27] by *Jock* [10].
2nd d. *White Bag* [5] Imported in 1853.

[139] JOCK O' BOMBIE.

Calved July 28th, 1868.

Bred by and the property of Mr. Thomas McCrae, Guelph, Co. Wellington, Ont.

Sire Dred [15].
Dam Lady Kenmure [140] by *Jock* [10].
2nd d. Pocahontas [60] by Marquis [19].
3rd d. Bonny [27] by Black Jock [18].
4th d. *White Bag* [5] Imported in 1853.

[143] DUCKIESTON.

Calved June 18th, 1867.

Bred by Mr. Thomas McCrae, Guelph, Co. Wellington, Ont.; the property of Mr. A. Quarry, Guelph, Co. Wellington, Ont.

Sire Dred [15].
Dam *Newbie Lass* [75] Imported.

[144] ROBIN HOOD.

Calved in 1866.

Bred by Mr. Wm. Hood, Guelph, Co. Wellington, Ont.; the property of Mr. H. W. Peterson, Hawkesville, Ont.

Sire *Pride of the Speed* [159].
Dam *Sall* [14] Imported in 1856.

BULLS.

[148]   BLACK DOUGLAS.

Calved May 24th, 1867.

Bred by Mr. Thomas McCrae, Guelph, Co. Wellington, Ont.; the property of Mr. John McCullock, Port Elgin, Ont.

Sire Dred [15].
Dam Pocahontas [60] by Marquis [19].
2nd d. Bonny [27] by *Jock* [10].
3rd d. *White Bag* [5] Imported in 1854.

[150]   LORD NAPIER.

Calved March 19th, 1866.

Bred by and the property of Mr. John Vassie, Ancaster, Co. Wentworth, Ont.

Sire William Wallace [67].
Dam Victoria [150] by *Black Prince* [152].
2nd d. *Beauty* [151] by *Peter* (116).

[157]   PRINCE OSCAR.

Calved February 24th, 1867.

Bred by and the property of Mr. John Coleman, West Flamboro', Co. Wentworth, Ont.

Sire Sir William Wallace [138].
Dam Jennie [152] by Garabaldi [130].
2nd d. Victoria [150] by *Black Prince* [152].
3rd d. *Beauty* [151] by *Peter* (116).

[152]   *BLACK PRINCE.*

Imported about 1857.

[154]   HARD FORTUNE.

Calved April 1st, 1864.

Bred by Mr. James Graham, Vaughan, Co. York, Ont.; the property of Mr. Arthur McNeil, Vaughan, Co. York, Ont.

Sire Dred [15].
Dam Bonny 2nd [74] by *Jock* [10].
2nd d. *Chloe* [4] Imported in 1853.

[155] PRINCE.

Calved March 20th, 1868.

Bred by and the property of Mr. Arthur McNeil, Vaughan, Co. York, Ont.

Sire *Dryfe* [222].
Dam Bonny 2nd [74] by *Jock* [10].
2nd d. *Chloe* [4] Imported in 1853.

[156] ROBIN HOOD.

Calved April 6th, 1868.

Bred by and the property of Mr. Arthur McNeil, Vaughan, Co. York, Ont.

Sire Hardfortune [154].
Dam Jenny Lind [32] by Marquis [19].
2nd d. *White Bag* [5] Imported in 1853.

[159] *PRIDE OF THE SPEED.*

Calved in 1863.

The property of Mr. Thomas McCrae, Guelph, Co. Wellington, Ont.; Imported in 1864 from the herd of Mr. Graham, Lockerby, Scotland.

[160] BLACK PRINCE.

Calved March 10th, 1868.

Bred by and the property of Mr. Thomas McCrae, Guelph, Co. Wellington, Ont.

Sire Dred [15].
Dam Queen of Beauty [95] by Black Jock [18].
2nd d. *Black Bess* [6] Imported in 1853.

[162] WELLINGTON *alias* ROB ROY.

Calved January 10th, 1869.

Bred by Mr. Thomas McCrae, Guelph, Co. Wellington, Ont.; the property of Mr. J. Giles, Boston, Mass., U. S.

Sire *Pride of the Speed* [159].
Dam *Polly Shaw* [174] Imported in 1861.

BULLS.

[163]   BERTIE.

Calved December 18th, 1866.

Bred by Mr. Thomas McCrae, Guelph, Co. Wellington, Ont.; the property of Mr. William Hurren, Mountsberg, Flamboro', Ont.

Sire Dred [15].
Dam Ebony [177] by *Jock* [10].
2nd d. *Blacky* [8] Imported in 1851.

[164]   YOUNG BLUCHER.

Calved April 16th, 1865.

Bred by and the property of Mr. James Charlton, Lobo, Co. Middlesex, Ont.

Sire Young Blucher [41].
Dam Victoria [150] by Black Prince [152].
2nd d. Shaw [21] by *The Pilgrim* [32] (32).
3rd d. *Sall* [14] Imported in 1856.

[165]   JOHNNY COPE.

Calved December 5th, 1869.

Bred by and the property of Mr. William Hood, Guelph, Co. Wellington, Ont.

Sire Our John [106].
Dam Nelly Grey [109] by Black Jock [18].
2nd. d. Jessie [49] by Black Jock [18].
3rd d. *Maggie* [48] by *Moss Trooper 2nd* [48] (151).

[166]   ROBIN.

Calved January 2nd, 1869.

Bred by and the property of Mr. William Hood, Guelph, Co. Wellington, Ont.

Sire *Pride of the Speed* [159].
Dam *Sall* [14] Imported in 1856.

[168]   LITTLE JOHN.

Calved November 1st, 1867.

Bred by and the property of Mr. Thomas McCrae, Guelph, Co. Wellington, Ont.

Sire Robin Hood [144].
Dam Maümee [92] by *Jock* [10].
2nd d. *Newbie Lass* [75] From the herd of Mr. James Beattie, Newbie House, Annan, Scotland.

[169] BLACK WELLINGTON.

Calved February 16th, 1870.

Bred by and the property of Mr. Joseph Charlton, Duncrief, Co. Middlesex, Ont.

Sire Young Blucher [41].
Dam Sall [170] by Robert Bruce [66].
2nd d. *Bonny* [1] Imported in 1853.

[175] KING TOM.

Calved March 20th, 1870.

Bred by and the property of Mr. Thomas McCrae, Guelph, Co. Wellington, Ont.

Sire *Pride of the Speed* [159].
Dam May Queen [143] by Honest Tom [98].
2nd d. Queen of Beauty [95] by Black Jock [18].
3rd d. *Black Bess* [6] Imported in 1853.

[176] ARGYLE.

Calved March 26th, 1870.

Bred by and the property of Mr. Thomas McCrae, Guelph, Co. Wellington, Ont.

Sire *Pride of the Speed* [159].
Dam *Polly Shaw* [174] Imported in 1861.

[184] SHOO-FLY.

Calved December 29th, 1870.

Bred by Mr. William Hood, Guelph, Co. Wellington, Ont.; the property of Mr. R. G. Hart, Lapeer, Michigan, U. S.

Sire Our John [106].
Dam Nelly Grey [109] by Black Jock [18].
2nd d. Jessie [49] by Black Jock [18].
3rd d. *Maggie* [48] by *Moss Trooper 2nd* [48] (151).

[190] PRINCE ALBERT.

Calved July 15th, 1859.

Bred by Mr. Joseph Jardine, Saltfleet, Co. Wentworth, Ont.; the property of Mr. Jeremiah Lyons, Dundas, Co. Wentworth, Ont.

Sire Moss Trooper [20].
Dam *Topsy* [13] Imported in 1856.

BULLS.

[196] JACOB BENNER'S BULL.

Age and breeder unknown.

[198] ROGER.

Calved February 2nd, 1870.

Bred by and the property of Mr. Arthur McNeil, Vaughan, Co. York, Ont.

Sire Black Prince [160].
Dam Woolwich Queen [96] by William Wallace [28].
2nd d. Black Bess [38] by Black Tom [37].
3rd d. *Bell* [9] Imported in 1853.

[199] JIM.

Calved April 4th, 1870.

Bred by and the property of Mr. Arthur McNeil, Vaughan, Co. York, Ont.

Sire Hardfortune [154].
Dam Susan [157] by *Dryfe* [222].
2nd d. Bess [125] by Marquis [19].
3rd d. *Chloe* [4] Imported in 1853.

[200] FRED.

Calved March 1st, 1871.

Bred by and the property of Mr. Arthur McNeil, Vaughan, Co. York, Ont.

Sire Hardfortune [154].
Dam Lizzie [114] by Dred [15].
2nd d. Bonny [74] by *Jock* [10].
3rd d. *Chloe* [4] Imported in 1853.

[204] WAVERLY.

Calved April 23rd, 1871.

Bred by and the property of Mr. Thos. McCrae, Guelph, Co. Wellington, Ont.

Sire *Pride of the Speed* [159]
Dam Minnie [178] by Dred [15].
2nd d. *Heather Bell* [12] Imported in 1856.

[207] JOCK 2ND.

Calved March 1st, 1867.

Bred by Mr. Arthur McNeil, Vaughan, Co. York, Ont.; the property of Mr. A. Harrison, Maugerville, New Brunswick.
Sire *Dryfe* [222].
Dam Black Bess [38] by Black Tom [37].
2nd d. *Bell* [9] Imported in 1853.

[214] WONDERFUL LAD.

Calved May, 1864.

Bred by Mr. Arthur McNeil, Vaughan, Co. York, Ont.; the property of Col. R. L. Denison, Dover Court, Toronto, Co. York, Ont.
Sire *Jock* [10].
Dam *Bell* [9] Imported in 1853.

[216] DOVER COURT.

Calved May 26th, 1866.

Bred by and the property of Col. R. L. Denison, Dover Court, Toronto, Co. York, Ont.
Sire Wonderful Lad [214].
Dam *Topsy* [213] Imported.

[218] LORD MONCK.

Calved March 6th, 1867.

Bred by and the property of Col. R. L. Denison, Dover Court, Toronto, Co. York, Ont.
Sire Cariboo [90].
Dam Mrs. Snell [105] by Black Jock [18].
2nd d. *Blacky* [8] Imported.

[219] DICK PETERSON.

Calved May 12th, 1865.

Bred by and the property of Col. R. L. Denison, Dover Court, Toronto, Co. York, Ont.
Sire Wonderful Lad [214].
Dam Salina [104] by Prince of Wales [57].
2nd d. *Helen* [103] Imported.

BULLS.

[220]   LIPPINCOTE.

Calved June 12th, 1867.

Bred by and the property of Col. R. L. Denison, Dover Court, Toronto, Co. York, Ont.

Sire Wonderful Lad [214].
Dam *Topsy* [213] Imported.

[222]   *DRYFE.*

Calved July, 1862.

Imported in 1863 by Mr. John Fleming, Vaughan, Co. York, Ont., from the herd of Mr. Graham, of Shaw Dryfe, Dumfries, Scotland.

[232]   PRINCE LEE BOO.

Calved November 16th, 1868.

Bred by Mr. James Graham, Vaughan, Co. York, Ont.; the property of Mr. Wm. Hood, Guelph, Co. Wellington, Ont.

Sire Jock [45].
Dam Woolwich Queen [96] by William Wallace [28].
2nd d. Black Bess [36] by Black Tom [37].
3rd d. *Bell* [9] Imported in 1853.

[238]   HONEST TOM 2ND.

Calved October 26th, 1871.

Bred by and the property of Mr. Wm. Hood, Guelph, Co. Wellington, Ont.

Sire Our John [106].
Dam Sall [192] by Prince Albert [190].
2nd d. *Black Bess* [6] Imported in 1853.

[239]   PRINCE ALBERT 2ND.

Calved December 17th, 1870.

Bred by and the property of Mr. William Hood, Guelph, Co. Wellington, Ont.

Sire Our John [106].
Dam Empress [50] by *Jock* [10].
2nd d. *Beauty* [11] Imported in 1856.

[240] ROBIN HOOD 2ND.

Calved April 2nd, 1872.

Bred by and the property of Mr. William Hood, Guelph, Co. Wellington, Ont.

Sire Our John [106].
Dam Lucy [502] by Black Jock [18].
2nd d. Jet [343] by Prince Albert [190].
3rd d. *Black Bess* [6] Imported in 1853.

[242] IVANHOE.

Calved May 7th, 1872.

Bred by and the property of Mr. Thomas McCrae, Guelph, Co. Wellington, Ont.

Sire King Tom [175]
Dam Minnie [178] by Dred [15].
2nd d. *Heather Bell* [12] Imported in 1856.

[245] YOUNG MOSS TROOPER.

Calved October 25th, 1871.

Bred by and the property of Mr. Joseph Charlton, Duncrief, Co. Middlesex, Ont.

Sire Young Blucher [41].
Dam Black Rachel [172] by Young Blucher [41].
2nd d. Sall [170] by Robert Bruce [66].
3rd d. *Bonny* [1] Imported in 1853.

[252] BOB.

Calved February 1st, 1872.

Bred by Mr. Arthur McNeil, Vaughan, Co. York, Ont. ; the property of Mr. Wm. Hood, Guelph, Co. Wellington, Ont.

Sire Hardfortune [154].
Dam Susan [15] by *Dryfe* [222].
2nd d. Bess [125] by Marquis [19].
3rd d. *Chloe* [4] Imported in 1853.

BULLS.

[253]   BLACK JOCK 2ND.

Calved March 17th, 1873.

Bred by and the property of Mr. Wm. Hood, Guelph, Co. Wellington, Ont.

Sire Robin [166].
Dam Nelly Grey [109] by Black Jock [18].
2nd d. Jessie [49] by Black Jock [18].
3rd d. *Maggie* [48] by *Moss Trooper 2nd* [48] (151).

[254]   JOCK A DINK.

Calved May 1st, 1878.

Bred by and the property of Mr. Wm. McCrae, Manor Bank, Guelph, Co. Wellington, Ont.

Sire Major Gray [273].
Dam Lady Dufferin [275] by King Tom [175].
2nd d. Mary Hay [147] by Dred [15].
3rd d. *Polly Shaw* [174] Imported.

[255]   HARDFORTUNE 2ND.

Calved May 16th, 1873.

Bred by and the property of Mr. Wm. Hood, Guelph, Co. Wellington, Ont.

Sire Robin [166].
Dam Lady Isabella [100] by Donald [123].
2nd d. *Chloe* [4] Imported in 1853.

[257]   LORD CECIL.

Calved January 22nd, 1873.

Bred by and the property of Mr. Thomas McCrae, Guelph, Co. Wellington, Ont.

Sire King Tom [175].
Dam Grace Darling [145] by Dred [15].
2nd d. Jessie [49] by Black Jock [18].
3rd d. *Maggie* [48] by *Moss Trooper 2nd* [48] (151).

[256] ROBIN HOOD 3RD.

Calved March 30th, 1873.

Bred by and the property of Mr. William Hood, Guelph, Co. Wellington, Ont.

Sire Robin [166].
Dam Lily Dale [233] by *Prince Oscar* [151].
2nd d. Jennie [152] by *Garibaldi* [130].
3rd d. Victoria [150] by *Black Prince* [152].
4th d. *Beauty* [151] by *Peter* (116).

[263] THE WARDEN.

Calved November 28th, 1875.

Bred by and the property of Mr. Thomas McCrae, Guelph, Co. Wellington, Ont.

Sire Lord Kenmure [271].
Dam Maggie Laidlaw [382] by *Pride of the Speed* [159].
2nd d. Maggie [120] by *Jock* [10].
3rd d. *Polly Shaw* [174] Imported.

[264] WELLINGTON BOY.

Calved March 23rd, 1873.

Bred by Mr. Wm. Hood, Guelph, Co. Wellington, Ont. ; the property of Captain Jack Whittaker, Oconomewoc, Wis., U. S.

Sire Robin [166].
Dam Sall [192] by Prince Albert [190].
2nd d. *Black Bess* [6] Imported.

[265] JOHNNY COPE.

Calved March 20th, 1874.

Bred by Mr. Wm. Hood, Guelph, Co. Wellington, Ont. ; the property of Mr. Peter Davy, Ashippun, Co. Dodge, Wis., U. S.

Sire Roger [198].
Dam Galloway Lass [234] by *Pride of the Speed* [159].
2nd d. Maumee [92] by *Jock* [10].
3rd d. *Newbie Lass* [75] Imported.

BULLS.

[266]   DUFFERIN.

Calved June 20th, 1873.

Bred by and the property of Mr. Thomas McCrae, Guelph, Co. Wellington, Ont.

Sire King Tom [175].
Dam Lady Kenmure [140] by *Jock* [10].
2nd d. Pocahontas [60] by Marquis [19].
3rd d. Bonny [27] by Black Jock [18].
4th d. *White Bag* [5] Imported.

[267]   SIR GARNET.

Calved January 3rd, 1875.

Bred by and the property of Mr. Thomas McCrae, Guelph, Co. Wellington, Ont.

Sire Lord Kenmure [271].
Dam Minnie [178] by Dred [15].
2nd d. *Heather Bell* [12] Imported.

[271]   LORD KENMURE.

Calved July 20th, 1870.

Bred by and the property of Mr. Thomas McCrae, Guelph, Co. Wellington, Ont.

Sire *Pride of the Speed* [159].
Dam *Newbie Lass* [75] Imported.

[272]   CLANDEBOYE.

Calved December 3rd, 1873.

Bred by and the property of Mr. Thos. McCrae, Guelph, Co. Wellington, Ont.

Sire Lord Kenmure [271].
Dam Mary Hay [147] by Dred [15].
2nd d. *Polly Shaw* [74] Imported.

[273]   MAJOR GRAY.

Calved March 31st, 1874.

Bred by and the property of Mr. Thos. McCrae, Guelph, Co. Wellington, Ont.

Sire Lord Kenmure [271].
Dam Mary Gray [183] by Bonnie Dundee [136].
2nd d. Mary Hay [147] by Dred [15].
3rd d. *Polly Shaw* [74] Imported.

[279] BLACK TOM.

Calved March 10th, 1867.

Bred by and the property of Mr. W. H. Peterson, Hawkesville, Co. Waterloo.

Sire Honest Tom [98].
Dam Black Jane [210] by Jacob Benner's Bull [196].
2nd d. *Black Bess* [6] Imported.

[280] SAM.

Calved February 20th, 1875.

Bred by and the property of Mr. Peter Davy, Ashippun, Co. Dodge, Wis., U.S.

Sire Bob [252].
Dam Jessie [257] by Black Tom [279].
2nd d. Wandering Nellie [277] by William Wallace [28].
3rd d. Miss Torrence [204] by William Wallace [28].
4th d. Black Jane [210] by Jacob Benner's Bull [196].
5th *Black Bess* [6] Imported.

[283] JOHNNY COPE 2ND.

Calved November 22nd, 1869.

Bred by Mr. William Hood, Guelph, Co. Wellington, Ont.; the property of Mr. R. G. Hart, Lapeer, Michigan, U.S.

Sire Our John [106].
Dam Lady Isabella [100] by Donald [123].
2nd d. *Chloe* [4] Imported.

[289] BUFFALO BILL.

Calved March 15th, 1873.

Bred by and the property of R. G. Hart, Lapeer, Michigan, U.S.

Sire Shoo Fly [184].
Dam Dairymaid [286] by *Dryfe* [222].
2nd d. Black Bess [38] by Black Tom [37].
3rd d. *Bell* [9] Imported.

BULLS. 27

[296] BADGER BOY.

Calved March 14th, 1876.

Bred by and the property of Mr. Peter Davy, Ashippun, Co. Dodge, Wisconsin, U. S.

Sire Wellington Boy [264].
Dam Fancy [202] by Hard Fortune [154].
2nd d. Woolwich Queen [96] by William Wallace [28].
3rd d. Black Bess [38] by Black Tom [37].
4th d. *Bell* [9] Imported.

[298] ROB ROY.

Calved February 3rd, 1877.

Bred by and the property of Mr. Peter Davy, Ashippun, Co. Dodge, Wisconsin, U. S.

Sire Wellington Boy [264].
Dam Jessie [257] by Black Tom [279].
2nd d. Wandering Nellie [277] by William Wallace [28].
3rd d. Miss Torrence [204] by William Wallace [28].
4th d. Black Jane [210] by Jacob Benner's Bull [196].
5th d. *Black Bess* [6] Imported.

[300] SCOTCH CHIEF.

Calved June 1st, 1877.

Bred by and the property of Mr. Peter Davy, Ashippun, Co. Dodge, Wis., U.S.

Sire Wellington Boy [264].
Dam Fancy [202] by Hardfortune [154].
2nd d. Woolwich Queen [96] by William Wallace [28].
3rd d. Black Bess [38] by Black Tom [37].
4th d. *Bell* [9] Imported.

[303] *YOUNG LOCHINVAR.*

Calved April, 1874.

Imported by and the property of Mr. Thomas McCrae, Guelph, Co. Wellington, Ont.

Sire *Kirkey*.
Dam *Rose* by *Son of Old Moss Trooper* (296.)

From the herd of Mr. John Underwood, Crofts, Kircudbright, Scotland.

[305] TOM SCOTT.

Calved March 10th, 1878.

Bred by and the property of Mr. Peter Davy, Ashippun, Co. Dodge, Wis., U.S.

Sire Sam [280].
Dam Mary [281] by Bob [252].
2nd d. Fancy [202] by Hardfortune [154].
3rd d. Woolwich Queen [96] by William Wallace [28].
4th d. Black Bess [38] by Black Tom [37].
5th d. *Bell* [9] Imported.

[308] CAPTAIN.

Calved February 10th, 1878.

Bred by and the property of Mr. Peter Davy, Ashippun, Co. Dodge, Wis., U.S.

Sire Sam [280].
Dam Jessie [257] by Black Tom [279].
2nd d. Wandering Nellie [277] by William Wallace [28].
3rd d. Miss Torrence [204] by William Wallace [28].
4th d. Black Jane [210] by Jacob Benner's Bull [196].
5th d. *Black Bess* [6] Imported.

[212] SNOWBALL.

Calved February 27th, 1878.

Bred by Mr. R. G. Hart, Lapeer, Mich., U.S.; the property of Mr. J. B. Sutherland, Petite Cote, Co. Essex, Ont.

Sire Buffalo Bill [289].
Dam Black Beauty [309] by Prince Albert 2nd [239].
2nd d. Black Rachel [294] by Shoo-Fly [184].
3rd d. Dairymaid [286] by *Dryfe* [222].
4th d. Black Bess [38] by Black Tom [37].
5th d. *Bell* [9] Imported.

313] JOHNNY SCOTT.

Calved February 15th, 1871.

Bred by Mr. J. N. Smith, Bath, Clinton Co., Mich., U.S.; the property of the Michigan State Agricultural College, Lansing, Mich., U.S.

Sire Our John [106].
Dam Maggie Lauder [110] by Black Jock [18].
2nd d. Empress [50] by *Jock* [10].
3rd d. *Beauty* [11] Imported.

BULLS.

[321]   HEANAN.

Calved February 20th, 1879.

Bred by and the property of Mr. Peter Davy, Ashippun, Co. Dodge, Wis., U.S.

Sire Sam [280].
Dam Sis [203] by Hardfortune [154].
2nd d. Susan [157] by *Dryfe* [222].
3rd d. Bess [127] by Marquis [19].
4th d. *Chloe* [4] Imported.

[325]   BADGER.

Calved March 18th, 1879.

Bred by and the property of Mr. Peter Davy, Ashippun, Co. Dodge, Wis., U.S.

Sire Sam [280].
Dam Fancy [202] by Hardfortune [154].
2nd d. Woolwich Queen [96] by William Wallace [28].
3rd d. Black Bess [38] by Black Tom [37].
4th d. *Bell* [9] Imported.

[327]   BILLY McNEIL.

Calved February 22nd, 1875.

Bred by and the property of Mr. J. N. Smith, Bath, Mich., U. S.

Sire Robin Hood [349].
Dam Rosa McNeil [345] by Hardfortune [154].
2nd d. Lizzie [114] by Dred [15].
3rd d. Bonny [74] by *Jock* [10].
4th d. *Chloe* [4] Imported.

[331]   DALRYMPLE.

Calved January 3rd, 1876.

Bred by and the property of Mr. Thomas McCrae, Guelph, Co. Wellington, Ont.

Sire Lord Kenmure [271].
Dam Mary Hay [147] by Dred [15].
2nd d. *Polly Shaw* [174] Imported.

[332] LORD WELLINGTON.

Calved November 10th, 1874.

Bred by Mr. Thomas McCrae, Guelph, Co. Wellington, Ont.; the property of Joseph Hickson, Montreal, Que.

Sire Lord Kenmure [271].
Dam Mary Hay [147] by Dred [15].
2nd d. *Polly Shaw* [174] Imported.

[333] DON BERNARDO.

Calved February 17th, 1877.

Bred by Mr. R. G. Hart, Lapeer, Michigan, U. S.; the property of Mr. W. B. Montgomery, Starkville, Miss., U. S.

Sire Buffalo Bill [289].
Victoria [290] by Shoo Fly [184].
2nd d. Black Bess 2nd [284] by Johnny Cope [283].
3rd d. Margaret [285] by *Dryfe* [222].
4th d. Black Bess [38] Black Tom [37].
5th d. *Bell* [9] Imported.

[334] LORD STARR.

Calved January 10th, 1876.

Bred by and the property of Mr. Thomas McCrae, Guelph, Co. Wellington, Ont.

Sire Lord Kenmure [271].
Dam Mary Gray [183] by Bonnie Dundee [136].
2nd d. Mary Hay [147] by Dred [15].
3rd d. *Polly Shaw* [174] Imported.

[335] CENTENNIAL.

Calved October 10th, 1876.

Bred by and the property of Mr. Geo. Hood, Guelph, Co. Wellington, Ont.

Sire Black Jock 2nd [253].
Dam Mary [111] by Our John [106].
2nd d. Empress [50] by *Jock* [10].
3rd d. *Beauty* [11] Imported.

BULLS.

[336] CULLY.

Calved March 7th, 1880.

Bred by and the property of Mr. Thomas McCrae, Guelph, Co. Wellington, Ont.

Sire Laird of Burleigh [516].
Dam Lady Dufferin [275] by King Tom [175].
2nd d. Mary Gray [183] by Bonnie Dundee [136].
3rd d. Mary Hay [147] by Dred [15].
4th d. *Polly Shaw* [174] Imported.

[337] ZULU.

Calved May 1st, 1879.

Bred by and the property of Mr. Thomas McCrae, Guelph, Co. Wellington, Ont.

Sire Lord Lochinvar [405].
Dam Lady Dufferin [275] by King Tom [175].
2nd d. Mary Gray [183] by Bonnie Dundee [136].
3rd d. Mary Hay [147] by Dred [15].
4th d. *Polly Shaw* [174] Imported.

[338] JOCK GRAY.

Calved February 1st, 1877.

Bred by and the property of Mr. Thomas McCrae, Guelph, Co. Wellington, Ont.

Sire Clandeboye [272].
Dam Mary Gray [183] by Bonnie Dundee [136].
2nd d. Mary Hay [147] by Dred [15].
3rd d. *Polly Shaw* [174] Imported.

[339] MACGREGOR.

Calved June 27th, 1877.

Bred by and the property of Mr. Thos. McCrae, Guelph, Co. Wellington, Ont.

Sire Major Gray [273].
Dam Minnie [178] by Dred [15].
2nd d. *Heather Bell* [12] Imported.

[340] SIR WILLIAM.

Calved May 15th, 1880.

Bred by and the property of Mr. Peter Davy, Ashippun, Co. Dodge, Wis., U.S.

Sire Sam [280].
Dam Dora Bell [297] by Wellington Boy [264].
2nd d. Sis [203] by Hardfortune [154].
3rd d. Susan [157] by *Dryfe* [222].
4th d. Bess [125] by Marquis [19].
5th d. *Chloe* [4] Imported.

[341] COMET.

Calved February 29th, 1880.

Bred by and the property of Mr. Peter Davy, Ashippun, Co. Dodge, Wis., U.S.

Sire Sam [280].
Dam Rose [306] by Sam [280].
2nd d. Sis [203] by Hardfortune [154].
3rd d. Susan [157] by *Dryfe* [222].
4th d. Bess [125] by Marquis [19].
5th d. *Chloe* [4] Imported.

[342] EMPEROR.

Calved March 2nd, 1880.

Bred by and the property of Mr. Peter Davy, Ashippun, Co. Dodge, Wis., U.S.

Sire Sam [280].
Dam Isabel [282] by Roger [198].
2nd d. Sis [203] by Hardfortune [154].
3rd d. Susan [157] by *Dryfe* [222].
4th d. Bess [125] by Marquis [19].
5th d. *Chloe* [4] Imported.

[346] NAPIER.

Calved April 27th, 1869.

Bred by Mr. Arthur McNeil, Vaughan, Co. York, Ont.; the property of Mr. J. N. Smith, Bath, Mich., U.S.

Sire Hardfortune [154].
Dam Young Bonny [39] by Black Tom [37].
2nd d. *Bell* [9] Imported.

BULLS.

[349] ROBIN HOOD.

Calved June 30th, 1871.

Bred by Mr. Wm. Hood, Guelph, Co. Wellington, Ont. ; the property of Mr. J. N. Smith, Bath, Mich., U.S.
    Sire Shoo-Fly [184].
    Dam Rosy Hood [188] by Our John [106].
    2nd d. Maggie Lauder [110] by Black Jock [18].
    3rd d. Empress [50] by *Jock* [10].
    4th d. *Beauty* [11] Imported.

[351] BILLY HOOD.

Calved April 20th, 1873.

Bred by Mr. J. N. Smith, Bath, Mich., U.S. ; the property of the Agricultural College, Kansas, U.S.
    Sire Robin Hood [349].
    Dam Maggie Lauder [110] by Black Jock [18].
    2nd d. Empress [50] by *Jock* [10].
    3rd d. *Beauty* [11] Imported.

[354] JOHNNY HOOD.

Calved August 6th, 1876.

Bred by Mr. J. N. Smith, Bath, Mich., U. S. ; the property of Mr. R. B. Caruss, St. Johns, Mich., U. S.
    Sire Robin Hood [349].
    Dam Rosy Hood [188] by Our John [106].
    2nd d. Maggie Lauder [110] by Black Jock [18].
    3rd d. Empress [50] by *Jock* [10].
    4th d. *Beauty* [11] Imported.

[355] BLACKWOOD.

Calved June 17th, 1877.

Bred by and the property of Mr. Thomas McCrae, Guelph, Co. Wellington, Ont.
    Sire Major Gray [273].
    Dam Lady Dufferin [275] by King Tom [175].
    2nd d. Mary Gray [183] by Bonnie Dundee [136].
    3rd d. Mary Hay [147] by Dred [15].
    4th d. *Polly Shaw* [174] Imported.

[366] JOHNNY McNEIL.

Calved March 3rd, 1879.

Bred by and the property of Mr. R. B. Caruss, St. Johns, Mich., U.S.
Sire Johnny Hood [354].
Dam Rosa McNeil [345] by Hardfortune [154].
2nd d. Lizzie [114] by Dred [15].
3rd d. Bonny [74] by *Jock* [10].
4th d. *Chloe* [4] Imported.

[370] INDEPENDENCE.

Calved July 4th, 1880.

Bred by and the property of Mr. R. B. Caruss, St. Johns, Mich., U.S.
Sire Johnny Hood [354].
Dam Jet 3rd [358] by Robin Hood [349].
2nd d. Jet [343] by Prince Albert [190].
3rd d. *Black Bess* [6] Imported.

[371] NERO 2ND.

Calved May 2nd, 1880.

Bred by and the property of Mr. R. B. Caruss, St. Johns, Mich., U.S.
Sire Johnny Hood [354].
Dam Jet 2nd [352] by Robin Hood [349].
2nd d. Jet [343] by Prince Albert [190].
3rd d. *Black Bess* [6] Imported.

[373] NERO 3RD.

Calved April 10th, 1880.

Bred by Mr. J. N. Smith, Bath, Mich., U.S.; the property of Mr. J. McDaniels, Mich., U.S.
Sire Billy McNeil [327].
Dam Rosy Hood 2nd [359] by Billy McNeil [327].
2nd d. Rosy Hood [188] by Our John [166].
3rd d. Maggie Lauder [110] by Black Jock [18].
4th Empress [50] by *Jock* [10].
5th d. *Beauty* [11] Imported.

[375] BILL.

Calved January 15th, 1881.

Bred by and the property of Mr. Peter Davy, Ashippun, Dodge Co., Wis., U.S.

Sire Johnny Scott [313].
Dam Bonny Lass [295] by Wellington Boy [264].
2nd d. Jessie [267] by Black Tom [279].
3rd d. Wandering Nellie [277] by William Wallace [28].
4th d. Miss Torrence [204] by William Wallace [28].
5th d. Black Jane [210] by Jacob Benner's Bull [196].
6th d. *Black Bess* [6] Imported.

[379] SIR WALTER.

Calved April 10th, 1881.

Bred by and the property of Mr. Peter Davy, Ashippun, Dodge Co., Wis., U.S.

Sire Johnny Scott [313].
Dam Cherry [307] by Sam [280].
2nd d. Fancy [202] by Hardfortune [154].
3rd d. Woolwich Queen [96] by William Wallace [28].
4th d. Black Bess [38] by Black Tom [37].
5th d. *Bell* [9] Imported.

[380] ROB.

Calved March 5th, 1881.

Bred by and the property of Mr. Peter Davy, Ashippun, Dodge Co., Wis., U.S.

Sire Johnny Scott [313].
Dam Sis [203] by Hardfortune [154].
2nd d. Susan [157] by *Dryfe* [222].
3rd d. Bess [125] by Marquis [19].
4th d. *Chloe* [4] Imported.

[381]   BEN.
Calved February 17th, 1881.

Bred by Mr. Peter Davy, Ashippun, Dodge Co., Wis., U.S.; the property of Mr. J. E. Hurlburt, Eldora, Iowa, U.S.
Sire Johnny Scott [313].
Dam Jessie [257] by Black Tom [279].
2nd d. Wandering Nellie [277] by William Wallace [28].
3rd d. Miss Torrence [204] by William Wallace [28].
4th d. Black Jane [201] by Jacob Benner's Bull [196].
5th d. *Black Bess* [6] Imported.

[386]   ROBBIE BURNS.
Calved November 4th, 1879.

Bred by Mr. Thomas McCrae, Guelph, Co. Wellington, Ont.; the property of Mr. A. B. Mathews, Kansas City, Mo., U.S.
Sire Lord Lochinvar [405].
Dam Rose [385] by Black Jock 2nd [253].
2nd d. Galloway Lass [234] by *Pride of the Speed* [159].
3rd d. Maümee [92] by *Jock* [10].
4th *Newbie Lass* [75] Imported.

[387]   DOCTOR HORNBOOK.
Calved February 8th, 1881.

Bred by Mr. Thomas McCrae, Guelph, Co. Wellington, Ont.; the property of Mr. A. B. Mathews, Kansas City, Mo., U.S.
Sire Lord Douglas [406].
Dam Rose [385] by Black Jock 2nd [253].
2nd d. Galloway Lass [234] by *Pride of the Speed* [159].
3rd d. Maümee [92] by *Jock* [10].
4th d. *Newbie Lass* [75] Imported.

[390]   PETER STEWART.
Calved March, 7th 1881.

Bred by Mr. Thos. McCrae, Guelph, Co. Wellington, Ont.; the property of Mr. A. B. Mathews, Kansas City, Mo., U.S.
Sire Major Gray [273].
Dam Border Lass [384] by *Young Lochinvar* [303].
2nd d. Minnie [178] by Dred [15].
3rd d. *Heather Bell* [12] Imported.

[391] BLACKAMOOR.

Calved April 27th, 1878.

Bred by Mr. Thos. McCrae, Guelph, Co. Wellington, Ont.; the property of Mr. A. B. Mathews, Kansas City, Mo., U.S.
Sire Clandeboye [272].
Dam Annie Laurie 2nd [251] by Johnny Cope [165].
2nd d. May Queen [143] by Honest Tom [98].
3rd d. Queen of Beauty [95] by Black Jock [18].
4th d. *Black Bess* [6] Imported.

[392] CETEWAYO.

Calved November 1st, 1879.

Bred by Mr. Thos. McCrae, Guelph, Co. Wellington, Ont.; the property of Mr. A. B. Mathews, Kansas City, Mo., U.S.
Sire Lord Douglas [406].
Dam Dagmar [153] by Black Jock [18].
2nd d. Jennie [152] by Garibaldi [130].
3rd d. Victoria [150] by *Black Prince* [152].
4th d. *Beauty* [151] by *Peter* (116).

[393] BISMARCK.

Calved July 4th, 1880.

Bred by Mr. Thos. McCrae, Guelph, Co. Wellington, Ont.; the property of Mr. A. B. Mathews, Kansas City, Mo., U.S.
Sire Lord Lochinvar [405].
Dam Lady Heron [180] by Sir John A. [137].
2nd d. Minnie [178] by Dred [15].
3rd d. *Heather Bell* [12] Imported.

[394] PRINCE.

Calved July 30th, 1878.

Bred by Mr. Thos. McCrae, Guelph, Co. Wellington, Ont.; the property of Mr. A. B. Mathews, Kansas City, Mo., U.S.
Sire Clandeboye [272].
Dam Dagmar [153] by Black Jock [18].
2nd d. Jennie [152] by Garibaldi [130].
3rd d. Victoria [150] by *Black Prince* [152].
4th d. *Beauty* [151] by *Peter* (116).

[395] TOM.

Calved March 20th, 1881.

Bred by Mr. R. B. Caruss, St. Johns, Michigan, U.S.; the property of Mr A. B. Mathews, Kansas City, Mo., U.S.

Sire Johnny Hood [354].
Dam Margaret [344] by Hardfortune [154].
2nd d. Lizzie [114] by Dred [15].
3rd d. Bonny [74] by *Jock* [10].
4th d. *Chloe* [4] Imported.

[404] WELLINGTON 2ND.

Calved April 23rd, 1875.

Bred by Mr. Wm. Hood, Guelph, Co. Wellington, Ont.; the property of Mr. A. B. Mathews, Kansas City, Mo., U.S.

Sire Bob [252].
Dam Dagmar [153] by Black Jock [18].
2nd d. Jennie [152] by Garibaldi [130].
3rd d. Victoria [150] by *Black Prince* [152].
4th d. *Beauty* [151] by *Peter* (116).

[405] LORD LOCHINVAR.

Calved July 17th, 1876.

Bred by Mr. Thomas McCrae, Guelph, Co. Wellington, Ont; the property of Mr. R. B. Caruss, St. Johns, Mich., U.S.

Sire *Young Lochinvar* [303].
Dam Queen Mab [206] by *Pride of the Speed* [159].
2nd d. May Queen [143] by Honest Tom [98].
3rd d. Queen of Beauty [95] by Black Jock [18].
4th d. Black Bess [6] Imported.

[406] LORD DOUGLAS.

Calved 1st April, 1876.

Bred by and the property of Mr. Thomas McCrae, Guelph, Co. Wellington, Ont.

Sire Lord Kenmure [271].
Dam Lady Heron [180] by Sir John A. [137].
2nd d. Minnie [178] by Dred [15].
3rd d. *Heather Bell* [12] Imported.

BULLS.

[410] BOB WOOLEY.

Calved May 22nd, 1877.

Bred by Mr. Anderson De Witt, Washington Court House, Ohio, U.S.; the property of Mr. A. B. Mathews, Kansas City, Mo., U.S.
Sire Modock [640].
Dam Margaret [344] by Hardfortune [154].
2nd d. Lizzie [114] by Dred [15].
3rd d. Bonny [74] by *Jock* [10].
4th d. *Chloe* [4] Imported.

[411] YOUNG SOLWAY.

Calved July 2nd, 1878.

Bred by and the property of Mr. Wm. McCrae, Manor Bank, Guelph, Co. Wellington, Ont.
Sire Major Gray [273].
Dam Lady Heron [180] by Sir John A. [45].
2nd Minnie [178] by Dred [15].
3rd d. *Heather Bell* [12] Imported.

[412] SCOTT.

Calved July 15th, 1880.

Bred by Mr. J. N. Smith, Bath, Mich., U.S.; the property of Messrs. Scott & Thrall, Co. Greenwood, Kansas, U.S.
Sire Billy McNeil [327].
Dam Rosy Hood [188] by Our John [106].
2nd d. Maggie Lauder [110] by Black Jock [18].
3rd d. Empress [50] by *Jock* [10].
4th d. *Beauty* [11] Imported.

[413] GOVERNOR ST. JOHN.

Calved April 23rd, 1881.

Bred by Mr. R. B. Caruss, St. Johns, Co. Clinton, Mich., U.S.; the property of Mr. M. R. Platt, Kansas City, Mo., U.S.
Sire Johnny Hood [354].
Dam Jet 3rd [358] by Robin Hood [349].
2nd d. Jet [343] by Prince Albert [190].
3rd d. *Black Bess* [6] Imported.

[425] UNCLE JOE.

Calved April 10th, 1880.

Bred by Mr. Jacob Weiste, Co. Lancaster, Pa., U.S.; the property of Mr. M. R. Platt, Kansas City, Mo., U.S.

Sire Wellington Boy [432].
Dam Lady Gordon [419] by Lord Kenmure [271].
2nd d. Lady Dufferin [275] by King Tom [175].
3rd d. Mary Gray [183] by Bonnie Dundee [136].
4th d. Mary Hay [147] by Dred [15].
5th d. *Polly Shaw* [174] Imported.

[426] GALLOWAY BOY.

Calved April 5th, 1879.

Bred by Mr. Jacob Weiste, Co. Lancaster, Pa., U.S.; the property of Mr. M. R. Platt, Kansas City, Mo., U.S.

Sire Wellington Boy [432].
Dam Sleeping Maggie [433] by Bob [278].
2nd d. May Queen [143] by Honest Tom [98].
3rd d. Queen of Beauty [95] by Black Jock [18].
4th d. *Black Bess* [6] Imported.

[427] ABERDEEN.

Calved July 2nd, 1879.

Bred by Mr. Jacob Weiste, Co. Lancaster, Pa.., U.S.; the property of Mr. M. R. Platt, Kansas City, Mo., U.S.

Sire Wellington Boy [432].
Dam Lady Hamilton [274] by Lord Kenmure [271].
2nd d. Lady Kenmure [140] by *Jock* [10].
3rd d. Pocahontas [60] by Marquis [19].
4th d. Bonny [27] by Black Jock [19].
5th d. *White Bag* [4] Imported.

BULLS.

[428] KING JACOB.

Calved April 2nd, 1878.

Bred by Mr. Jacob Weiste, Co. Lancaster, Pa., U.S.; the property of Mr. M. R. Platt, Kansas City, Mo., U.S.

Sire Wellington Boy [432].
Dam Lady Gordon [419] by Lord Kenmure [271].
2nd d. Lady Dufferin [275] by King Tom [175].
3rd d. Mary Gray [183] by Bonnie Dundee [136].
4th d. Mary Hay [147] by Dred [15].
5th d. *Polly Shaw* [175] Imported.

[429] JOHN BROWN.

Calved June 23rd, 1881.

Bred by and the property of Mr. M. R. Platt, Kansas City, Mo., U.S.

Sire Billy McNeil [327].
Dam Jettie 3rd [361] by Billy McNeil [327].
2nd d. Jettie [348] by Our John [106].
3rd d. Jet [343] by Prince Albert [190].
4th d. *Black Bess* [6] Imported.

[430] JIM CROW.

Calved December 5th, 1879.

Bred by the Nebraska State University, U.S.; the property of Mr. M. R. Platt, Kansas City, Mo., U.S.

Sire McNeil [443].
Dam Lassie [445] by McNeil [443].
2nd d. Snowflake [315] by Johnny Scott [313].
3rd d. Snowball [314] by Victor [122].
4th d. Blooming Heather [80] by Dred [15].
5th d. *Blooming Heather* [53] by *Moss Trooper 2nd* [48] (151).
6th d. *Mary* by *Fergy* (19).
7th d. *Jane* by *Bob*.

[431] *BOTHWELL.*

Calved October, 25th, 1880.

The property of Mr. Peter Davy, Oconomewoc, Wis., U.S.; imported by Mr. Simon Beattie in 1881.

Sire in Scotland; owned by Mr. James Cunningham, Tarbreoch.
Dam owned by the Earl of Galloway, in Scotland.

[432] WELLINGTON BOY.

Calved February 3rd, 1876.

Bred by and the property of Mr. Wm. Hood, Guelph, Co. Wellington Ont.

Sire Black Jock 2nd [253].
Dam Galloway Lass [234] by *Pride of the Speed* [159].
2nd d. Maümee [92] by *Jock* [10].
3rd d. *Newbie Lass* [75] Imported.

[435] NELSON.

Calved June 20th, 1875.

Bred by and the property of Mr. William Hood, Guelph, Co. Wellington, Ont.

Sire Bob [278].
Dam Lily Dale [233] by Prince Oscar [151].
2nd d. Jennie [152] by Garibaldi [130].
3rd d. Victoria [150] by *Black Prince* [152].
4th d. *Beauty* [151] by *Peter* (119).

[437] DUNCAN.

Calved in 1868.

Bred by Mr. Arthur McNeil, Vaughan, Co. York, Ont.; the property of Mr. Peter Davy, Ashippun, Co. Dodge, Wis., U. S.

Sire *Dryfe* [222].
Dam Young Bonny [39] by Black Tom [37].
2nd d. *Bell* [9] Imported.

[439] SILVER KING.

Calved July 28th, 1881.

Bred by Mr. Jacob Weiste, Co. Lancaster, Pa., U. S.; the property of Mr. M. R. Platt, Kansas City, Mo., U. S.

Sire Aberdeen [427].
Dam Queen [418] by Wellington Boy [432].
2nd d. Sleeping Maggie [433] by Bob [252].
3rd d. May Queen [143] by Honest Tom [98].
4th d. Queen of Beauty [95] by Black Jock [18].
5th d. *Black Bess* [6] Imported.

BULLS.

[440] ### UNCLE SAM.

Calved May 15th, 1880.

Bred by Mr. Jacob Weiste, Co. Lancaster, Pa., U. S.; the property of Mr. M. R. Platt, Kansas City, Mo., U. S.
Sire Wellington Boy [432].
Dam Sleeping Maggie [433] by Bob [252].
2nd d. May Queen [143] by Honest Tom [98].
3rd d. Queen of Beauty [95] by Black Jock [18].
4th d. *Black Bess* [6] Imported.

[441] ### HOPEWELL.

Calved April 3rd, 1879.

Bred by and the property of Mr. Daniel Steinmety, Schoeneck, Pa., U.S.
Sire Wellington Boy [432].
Dam Lady Gordon [419] by Lord Kenmure [271].
2nd d. Lady Dufferin [275] by King Tom [175].
3rd d. Mary Gray [183] by Bonnie Dundee [136].
4th d. Mary Hay [147] by Dred [15].
5th d. *Polly Shaw* [174] Imported.

[443] ### McNEIL.

Calved May 2nd, 1873.

Bred by Mr. J. N. Smith, Bath, Mich., U.S.; the property of the State University of Nebraska, U.S.
Sire Napier [346].
Dam Rosa McNeil [345] by Hardfortune [154].
2nd d. Lizzie [114] by Dred [15].
3rd d. Bonny [74] by *Jock* [10].
4th d. *Chloe* [4] Imported.

[447 ### LORNE.

Calved April 10th, 1879.

Bred by and the property of Mr. Thos. McCrae, Guelph, Co. Wellington, Ont.
Sire Clandeboye [272].
Dam Galloway Lass [234] by *Pride of the Speed* [159].
2nd d. Maümee [92] by *Jock* [10].
Newbie Lass [75] Imported.

[449] BLACK JOCK.

Calved February 26th, 1881.

Bred by Mr. Wm. Edie, Dunnville, Ont.; the property of Mr. D. Saurn, Olin, Co. Jones, Iowa, U.S.

Sire Lorne [447].
Dam Jenny Lind [448] by Black Jock 2nd [253].
2nd d. Gipsy Queen [194] by Our John [106].
3rd d. Dido [88] by Victor [122].
4th d. Magnolia [847] by Black Jock [18].
5th d. Galloway Lass [42] by *Jock* [10].
6th d. *Jet* [2] Imported.

[455] SCOTTISH CHIEF.

Calved 5th March, 1876.

Bred by Mr. Philo Lasher, Coffeysburg, Co. Davies, Mo., U.S.; the property of Mr. M. R. Platt, Kansas City, Mo., U.S.

Sire Wellington Boy [264].
Dam Snowdrift [444] by Napier [346].
2nd d. Snowball [314] by Victor [122].
3rd Blooming Heather 2nd [80] by Dred [15].
4th d. *Blooming Heather* [53] by *Moss Trooper 2nd* [48] (151).
5th d. *Mary* by *Fergy* (19).
6th d. *Jane* by *Bob*.

[456] JOHNNY COPE 2ND.

Calved April 20th, 1879.

Bred by Mr. Philo Lasher, Coffeysburg, Co. Davies, Mo., U.S.; the property of Mr. M. R. Platt, Kansas City, Mo., U.S.

Sire Johnny Cope [265].
Dam Snowdrift [444] by Napier [346].
2nd d. Snowball [314] by Victor [122].
3rd d. Blooming Heather 2nd [80] by Dred [15].
4th d. *Blooming Heather* [53] by *Moss Trooper 2nd* [48] (151).
5th d. *Mary* by *Fergy* (19).
6th d. *Jane* by *Bob*.

BULLS.

[457]   CHARLEY COPE.

Calved April 10th, 1878.

Bred by Mr. Philo Lasher, Coffeysburg, Co. Davies, Mo., U.S.; the property of Mr. Wm. Clelland, Martinville, Co. Harrison, Mo., U.S.

Sire Johnny Cope [265].
Dam Snowdrift [444] by Napier [346].
2nd d. Snowball [314] by Victor [122].
3rd d. Blooming Heather 2nd [80] by Dred [15].
4th d. *Blooming Heather* [53] by *Moss Trooper 2nd* [48] (151).
5th d. *Mary* by *Fergy* (19).
6th d. *Jane* by *Bob*.

[458]   PETER COPE.

Calved August 1st, 1878.

Bred by Mr. Philo Lasher, Coffeysburg, Co. Davies, Mo., U.S.; the property of Mr. Wesley W. Mock, Martinville, Co. Harrison, Mo., U.S.

Sire Johnny Cope [265].
Dam Blue Bell [438] by Duncan [437].
2nd d. Belle Mahone [196] by Our John [106].
3rd d. Nellie Grey [109] by Black Jock [18].
4th d. Jessie [49] by Black Jock [18].
5th *Maggie* [48] by *Moss Trooper 2nd* [48] (151).

[470]   *MACLEOD OF DRUMLANRIG* (1564).

Calved January 2nd, 1880.

Bred by Duke of Buccleuch, K.G., Scotland; imported by Mr. Thos. McCrae, Guelph, Co. Wellington, Ont.; and the property of Mr. F. McHardy, Emporia, Kansas, U.S.

Sire *Stanley of Drumlanrig* (1348).
Dam *Harriet 4th of Drumlanrig* (2622) by *Black Prince of Drumlanrig* (546).
2nd d. *Harriet of Drumlanrig* (1636) by *Heir-at-Law* (815).
3rd d. *Nancy of Elrig* (2699) by *Tawny*.
4th d. *Kate* from the herd of Mr. Wm. Routledge, Port William, Wigtonshire.

[471]   *MACLEOD OF DARE.*

Calved 9th April, 1882.

Imported by Mr. Thos. McCrae, Guelph, Co. Wellington, Ont.; bred by Mr. James Cunningham, Tarbreoch, Dalbeattie, Scotland; and the property of Mr. F. McHardy, Emporia, Kansas, U.S.

Sire *Macleod of Drumlanrig* [470] (1564).
Dam *Countess of Galloway* [168] by *Scottish Borderer* (669).

[472]   *THE EARL.*

Calved 5th January, 1882.

Imported by Mr. Thomas McCrae, Guelph, Co. Wellington, Ont.; bred by Mr. John Graham, Kirkconnel, Castle Douglas, Scotland; and the property of Mr. F. McHardy, Emporia, Kansas, U.S.

Sire *Earlston* [478].
Dam *Primrose of Kirkconnel* [463] (3317).
2nd d. *Maggie*, from the herd of Mr. Potts, Bewcastle, Cumberland.

[473]   *DALBEATTIE.*

Calved 8th March, 1882.

Bred by Mr. J. Biggar & Sons, Dalbeattie, Scotland; imported by Mr. Thos. McCrae, Guelph, Co. Wellington, Ont.; the property of Mr. F. McHardy, Emporia, Kansas, U.S.

Sire *Knowsley* (1279).
Dam *Jane Seton 2nd* [467] (3787) by *Lord of Nithsdale* (616).
2nd d. *Jane Seton* (2697) by *Kirkey*.

[474]   ROYAL CHARLIE.

Calved November 25th, 1879.

Bred by Mr. Wm. McCrae, Manor Bank, Guelph, Co. Wellington, Ont.; the property of Mr. F. McHardy, Emporia, Kansas, U.S.

Sire Lord Lochinvar [405].
Dam Mary [111] by Our John [106].
2nd d. Empress [50] by *Jock* [10].
3rd d. *Beauty* [11] Imported.

BULLS.

[475] FRED.

Calved March 28th, 1875.

Bred by and the property of Mr. William Hood, Guelph, Co. Wellington, Ont.

Sire Bob [252].
Dam Black Sall [223] by Black Tom [279].
2nd d. White Bag [220] by Moss Trooper 4th [493].
3rd d. Black Swan [219] by Prince Albert [190].
4th d. Black Jessie [218] by *Jock* [10].
5th d. *Sall* [14] Imported.

[478] *EARLSTON.*

Calved about 1878.

Bred by and the property of Sir Wm. Gordon, Earlston, Scotland.

Sire *Larriston* (1030).
Dam *Esther of Tarbreoch* (2849).

[480] FRED 2ND.

Calved December 15th, 1878.

Bred by Mr. Christopher Wilson, Fergus, Co. Wellington, Ont.; the property of Mr. F. McHardy, Emporia, Kansas, U.S.

Sire Fred [475].
Dam Sall [485] by Roger [198].
2nd d. Lady Isabella [100] by Donald [123].
3rd d. *Chloe* [4] Imported.

[481] FRED 4TH.

Calved December 4th, 1881.

Bred by Mr. Christopher Wilson, Fergus, Co. Wellington, Ont.; the property of Mr. F. McHardy, Emporia, Kansas, U.S.

Sire Fred 2nd [480].
Dam Sall [485] by Roger [198].
2nd d. Lady Isabella [100] by Donald [123].
3rd d. *Chloe* [4] Imported.

[484]   PRINCE CHARLIE.
Calved April 15th, 1882.

Bred by Mr. Andrew Elliott, Galt, Co. Waterloo, Ont.; the property of Mr. F. McHardy, Emporia, Kansas, U.S.

Sire Royal Charlie [474].
Dam Mary [111] by Our John [106].
2nd d. Empress [50] by *Jock* [10].
3rd d. *Beauty* [11] Imported.

[489]   JOHNNY COPE 4TH.
Calved October 20th, 1881.

Bred by Mr. Philo Lasher, Coffeysburg, Mo., U.S.; the property of Mr. L. Miller, Marysville, Co. Nodaway, Mo., U.S.

Sire Johnny Cope 2nd [456].
Dam Lady Bell [450] by Johnny Cope [265].
2nd d. Blue Bell [428] by Duncan [437].
3rd d. Belle Mahone [196] by Our John [106].
4th d. Nellie Gray [109] by Black Jock [18].
5th d. Jessie [49] by Black Jock [18].
6th d. *Maggie* [48] by *Moss Trooper 2nd* [48] (151).

[490]   KING WILLIAM.
Calved October 24th, 1882.

Bred by and the property of Mr. Wm. Edie, Dunnville, Ont.

Sire Lorne [447].
Dam Gipsy Queen [194] by Our John [106].
2nd d. Dido [88] by Victor [122].
3rd d. Magnolia [84] by *Black Jock* [18].
4th d. Galloway Lass [42] by *Jock* [10].
5th d. *Jet* [2] Imported.

[491   BLACK DAN.
Calved May 23rd, 1882.

Bred by and the property of Palmer & Sons, Boscobel, Grant Co., Wis., U.S.

Sire Sam [280].
Dam Snovir [318] by Johnny Scott [313].
2nd d. Snowball [314] by Victor [122].
3rd d. Blooming Heather 2nd [80] by Dred [15].
4th d. *Blooming Heather* [53] by *Moss Trooper 2nd* [48] (151).
5th d. *Mary* by *Fergy* (19).
6th d. *Jane* by *Bob*.

BULLS. 49

[493] MOSS TROOPER 4TH.

Calved 1861.

Bred by Mr. Graham, Vaughan, Co. York, Ont. ; the property of Mr. Thos. McCrae, Guelph, Co. Wellington, Ont.

Sire *Moss Trooper* [74] (296).
Dam a pure cow imported by Mr. Graham, Vaughan.

[494] HERD LADDIE.

Calved June 17th, 1882.

Bred by and the property of Mr. Geo. Coleman, Howell, Co. Livingston, Mich., U.S.

Sire Shoo-Fly [184].
Dam Pride of the Dairy [497] by Hardfortune 2nd [255].
2nd d. Margaret [285] by *Dryfe* [222].
3rd d. Black Bess [38] by Black Tom [37].
4th d. *Bell* [9] Imported.

[501] ROB ROY.

Calved April 17th, 1878.

Bred by Mr. Geo. Coleman, Marion, Co. Livingston, Mich., U.S. ; the property of Mr. H. A. Disbrow, Atlantic, Iowa, U.S.

Sire Herd Laddie [494].
Dam Pride of the Dairy [497] by Hardfortune 2nd [255].
2nd d. Margaret [285] by *Dryfe* [222].
3rd d. Black Bess [38] by Black Tom [37].
4th d. *Bell* [9] Imported.

[508] LOCHINVAR, JR.

Calved June 18th, 1882.

Bred by and the property of Mr. R. B. Caruss, St. Johns, Mich., U.S

Sire Lord Lochinvar [405].
Dam Sally [365] by Johnny Hood [354].
2nd d. Jet 3rd [358] by Robin Hood [349].
3rd d. Jet [343] by Prince Albert [190].
4th d. *Black Bess* [6] Imported.

[516] LAIRD OF BURLEIGH.

Calved April 24th, 1878.

Bred by Mr. Wm. McCrae, Manor Bank, Guelph, Co. Wellington, Ont.; the property of Mr. Joseph Hickson, Montreal, Que.

Sire Major Gray [273].
Dam Jeanie Deans [243] by King Tom [175].
2nd d. Mary Hay [147] by Dred [15].
3rd d. *Polly Shaw* [174] Imported.

[519] DISRAELI.

Calved January 17th, 1877.

Bred by and the property of Mr. Wm. McCrae, Manor Bank, Guelph, Co. Wellington, Ont.

Sire Major Gray [273].
Dam Lady Heron [180] by Sir John A. [137].
2nd d. Minnie [178] by Dred [15].
3rd d. *Heather Bell* [12] Imported.

[520] PRIDE.

Calved April 31st, 1872.

Bred by and the property of Mr. Thomas McCrae, Guelph, Co. Wellington, Ont.

Sire King Tom [175].
Dam Maggie Laidlaw [382] by *Pride of the Speed* [159].
2nd d. Maggie [120] by Dred [15].
3rd d. *Polly Shaw* [174] Imported.

[521] LORD CHELMSFORD.

Calved November 10th, 1878.

Bred by Mr. Wm. McCrae, Manor Bank, Guelph, Co. Wellington, Ont.; the property of Mr. James Calvert, Dromore, Co. Grey, Ont.

Sire Lord Lochinvar [405].
Dam Maggie Laidlaw [382] by *Pride of the Speed* [159].
2nd d. Maggie [120] by Dred [15].
3rd d. *Polly Shaw* [174] Imported.

BULLS.

[529] *CHIEF TEMPLAR (1782).*

Calved January 10th, 1881.

Bred by Messrs. H. H. & I. Mitchelson, Lynes, Brampton, England; imported by and the property of Mr. Peter Davy, Oconomewoc, Wis., U.S.

Sire *The Miller o' Dee* (1373).
Dam *Beauty of Lynes* (4208) by *Theodore* (1778).
2nd d. *Kate* by *Lanercost*.

[530] *TROOPER.*

Calved May 10th, 1882.

Bred by Mr. John Millican, Wedholme House, Abbey Town, Carlisle; imported by and the property of Mr. Peter Davy, Oconomewoc, Wis., U.S.

Sire *Hazeldean* (1010).
Dam *Moss Rose of Wedholme* (2876) by *Peddar Lad* (1139).
3nd d. *Rose*.

[543] *CROMWELL 2nd (1759).*

Calved March 26th, 1881.

Bred by Mr. J. P. Chambers, Lousay House, Abbey Town, Carlisle; imported by and the property of Mr. Peter Davy, Oconomewoc, Wis., U.S.

Sire *Lochiel of Shaw* (1329).
Dam *Beauty 5th of Lousay* (2864) by *Black Bonnet* (575).
2nd d. *Beauty of Pelutho 2nd* (1365) by *Camerton* (542).
3rd d. *Beauty of Pelutho* (1358) by *Geordie 2nd of Green* (1135).

[546] ENCHANTER.

Calved March 1st, 1882.

Bred by and the property of Mr. Peter Davy, Oconomewoc, Wis., U.S.
Sire Johnny Cope [265].
Dam Isabel [282] by Roger [198].
2nd d. Sis [203] by Hardfortune [154].
3rd d. Susan [157] by *Dryfe* [222].
4th d. Bess [125] by Marquis [19].
5th d. *Chloe* [4] Imported.

[548] *ARABI BEY.*

Calved April 21st, 1882.

Bred by Mr. Andrew Milligan, Corwall, Port William, Scotland; imported by and the property of Mr. Thos. McCrae, Guelph, Co. Wellington, Ont.

Sire *Macleod of Tarbreoch* (1471).
Dam *Nellie of Corwall* (3885) by *Dominie Sampson* (1149).
2nd d. *Mary 3rd of Redcastle* (2938) by *Baron Douglas* (614).
3rd d. *Mary of Redcastle* (2926) from herd of Mr. Jas. Graham, Meikle Culloch, Dalbeattie, Scotland.

[549] *AUTUMN* (1698).

Calved February 21st, 1881.

Bred by Messrs. W. & J. Shennan, Balig, Scotland; imported by and the property of Mr. Thos. McCrae, Guelph, Co. Wellington, Ont.

Sire *Normandy* (1533).
Dam *Miller 6th* (3449) by *Uncle Tom of Balig* (1043).
2nd d. *Miller 3rd* (1390) by *Norman* (529).
3rd d. *Miller 2nd* (1350) by *The Goat* (527).
4th d. *Miller* (1339) by *Bob Burns* (235).
5th d. *Blacky of Corrahill* (375) from the herd of Mr. Wm. Bryce, Corrahill.

[550] *CHAPELTON.*

Calved June 16th, 1882.

Bred by Mr. Samuel Moffatt, Aird, Crossmichael, Scotland; imported by and the property of Mr. Thos. McCrae, Guelph, Co. Wellington, Ont.

Sire *Dick of Walton* (2195).
Dam *Maid of Dumfries* [573] (4974) by *Rob Roy of Chapelton* (1856).

[551] *ELRIG.*

Calved April 10th, 1881.

Bred by Mr. Wm. Routledge, Port William, Wigtonshire, Scotland; imported by and the property of Mr. Thomas McCrae, Guelph, Co. Wellington, Ont.

Sire *The Baron of Drumlanrig* (1158).
Dam *Maggie 2nd of Elrig* (3030) by *Scottish Borderer* (669).
2nd d. *Maggie of Elrig* (2701) by *Heir-at-Law* (815).
3rd. d. *Jean*, from the herd of Mr. Wm. Routledge, Port William.

BULLS.

[552] *LOFTY OF WATERSIDE.*

Calved September 10th, 1880.

Bred by Mr. Robert Smart, Dalbeattie, Scotland ; imported by Mr. Thos. McCrae, Guelph, Co. Wellington, Ont. ; and the property of Mr. J. N. Smith, Bath, Mich., U.S.

Sire *Loftus*.
Dam *Alma* from the herd of Mr. Nairn, Glenlair, one of the oldest bred Galloway herds in the district.

[553] *MACLEOD 2ND OF DRUMLANRIG* (1676).

Calved March 5th, 1881.

Bred by the Duke of Buccleuch, Scotland ; imported by and the property of Mr. Thos. McCrae, Guelph, Co. Wellington, Ont.

Sire *Stanley of Drumlanrig* (1348).
Dam *Harriet 4th of Drumlanrig* (2622) by *Black Prince of Drumlanrig* (546).
2nd d. *Harriet of Drumlanrig* (1636) by *Heir-at-Law* (815).
3rd d. *Nancy of Elrig* (2699) by *Tawny*.
4th d. *Kate* from the herd of Mr. Wm. Routledge, Port William, Wigtonshire, Scotland.

[554] *MOCHRUM.*

Calved February 22nd, 1881.

Bred by Mr. Wm. Routledge, Port William, Wigtonshire, Scotland imported by Mr. Thomas McCrae, Guelph, Co. Wellington, Ont. ; and the property of Col. N. Higginbotham, Guelph, Co. Wellington, Ont.

Sire *The Baron of Drumlanrig* (1158).
Dam *Mary 2nd of Elrig* (2705) by *Marquis of Elrig* (842).
2nd d. *Mary of Elrig* (2700) by *Heir-at-Law* (815).
3rd d. *Nancy of Elrig* (2699) by *Tawny*.
4th d. *Kate* from the herd of Mr. Wm. Routledge, Port William, Wigtonshire, Scotland.

[555]   SAM OF GARLIESTON (1610).
Calved March 12th, 1880.
Bred by the Earl of Galloway, Garlieston, Scotland; imported by Mr. Thomas McCrae, Guelph, Co. Wellington, Ont.; and the property of Mr. R. B. Caruss, St. Johns, Mich., U.S.
Sire *Scottish Borderer* (669).
Dam *Maggie* from the herd of the Earl of Galloway.

[587]   TOM *alias* TOMMY DODD.
Calved January 7th, 1880.
Bred by Mr. Geo. Coleman, Howell, Co. Livingston, Mich., U.S.; the property of Mr. Wm. Keith, Pittsford, Co. Hillsdale, Mich., U.S.
Sire Rob Roy [501].
Dam Chloe [584] by Herd Laddie [494].
2nd d. Pride of the Dairy [497] by Hardfortune 2nd [255].
3rd d. Margaret [285] by *Dryfe* [222].
4th d. Black Bess [38] by Black Tom [37].
5th d. *Bell* [9] Imported.

[588]   BLACK DIAMOND.
Calved January 1st, 1882.
Bred by Mr. Geo. Coleman, Howell, Co. Livingston, Mich., U.S.; the property of Mr. W. P. Darrow, Jefferson, Co. Hillsdale, Mich., U.S.
Sire Malcolm [614].
Dam Pride of the Dairy [497] by Hardfortune 2nd [255].
2nd d. Margaret [285] by *Dryfe* [222].
3rd d. Black Bess [38] by Black Tom [37].
4th d. *Bell* [9] Imported.

[589]   LAD.
Calved December 20th, 1880.
Bred by Mr. Wm. Keith, Pittsford, Co. Hillsdale, Mich., U.S.; the property of Mr. L. Romeyn Giddings, Downers Grove, Co. Du Page, Ill., U.S.
Sire Rob Roy [501].
Dam Sweet Aften [585] by Robin Hood [349].
2nd d. Chloe [584] by Herd Laddie [494].
3rd d. Pride of the Dairy [497] by Hardfortune 2nd [255].
4th d. Margaret [285] by *Dryfe* [222].
5th d. Black Bess [38] by Black Tom [37].
6th d. *Bell* [9] Imported.

BULLS.

[590] *BARON.*

Calved July, 1881.

Bred by Mr. James Cunningham, Tarbreoch, Scotland; imported by and the property of Mr. Simon Beattie, Annan, Scotland.

Sire *Knowsley* (1279).
Dam a pure Galloway cow from the herd of Mr. James Cunningham.

[591] *LOCHIEL.*

Calved May 6th, 1881.

Bred by Mr. George Greig, Dalry, Galloway, Scotland; imported by and the property of Mr. Simon Beattie, Annan, Scotland.

Sire *Maori Chief* [592] (1433).
Dam *Lilias of Milnmark* (3503) by *Marquis of Tarbreoch* (1049).
2nd d. *Betsie of Milnmark* (3132) by *Bogue*.
3rd d. *Canty* from the herd of Mr. Wilson, Troquhain, Balmaclellan, New Galloway, Scotland.

[592] *MAORI CHIEF* (1433).

Calved March 9th, 1878.

Bred by Mr. Gilbert Grieve, Minnydow, Dalbeattie, Scotland; imported by and the property of Mr. Simon Beattie, Annan, Scotland.

Sire *Larriston* (1030).
Dam *Maori of Minnydow* (3368) by *Marquis of Tarbreoch* (1049).
2nd d. *Juno 2nd of Tarbreoch* (1672) by *Havelock* (544).
3rd d. *Juno* (2960) by *Nelson* (1104).
4th d. *Nancy* from the herd of Mr. James Cunningham, Scotland.

[593] *PETER.*

Calved July, 1882.

Bred by Mr. Simon Beattie, Annan, Scotland; the property of Mr. Peter Davy, Oconomewoc, Wis., U.S.

Sire *Knowsley* (1279).
Dam *Alice Maxwell* [596] by *Tom of Crofts* by *Henry* (1127).
2nd d. a pure cow from the herd of Mr. Adam Thomson, Castle Douglas, Scotland.

[594]   SCOTIA OF TARBREOCH.

Calved April, 1879.

Bred by and the property of Sir Wm. Gordon, Bart, Kircudbright, Scotland; imported by Mr. Simon Beattie, Annan, Scotland.

Sire *Larriston* (1030).

Dam *Esther of Tarbreoch* (2849) from the herd of Mr. James Cunningham, Scotland.

[595].   YOUNG FRANKLIN.

Calved in 1880.

Bred by and the property of Mr. John Cannon, Dumfries, Scotland; imported by Mr. Simon Beattie, Annan, Scotland.

Sire *Franklin* (1292).

Dam *Beauty of Congeith* (2737) sire bred at Laggan.

2nd d. *Jenny of Congeith* (2735) from the herd of Mr. John Cannon, Dumfries, Scotland.

[613]   DE BARRE.

Calved June 27th, 1881.

Bred by Mr. R. De Barré Cunninghame, Hensol, Kirkcudbright, Scotland; imported by Mr. Thos. McCrae, Guelph, Co. Wellington, Ont.; and the property of Mr. Chas. M. Saxby, Freeport, Ill., U.S.

Sire *Sergeant Stanley* (1501).

Dam *Jenny of Hensol* (3699) by *The Dutchman* (1449).

2nd d. *Mullock of Hensol* (3301) from the herd of Mr. R. De Barré Cunninghame.

[614]   LORD MALCOLM.

Calved April 10th, 1880.

Bred by and the property of Mr. George Coleman, Howell, Co. Livingston, Mich., U.S.

Sire Rob Roy [501].

Dam Dinah [613] by Robin Hood [349].

2nd d. Chloe [584] by Herd Laddie [494].

3rd d. Pride of the Dairy [497] by Hardfortune 2nd [255].

4th d. Margaret [255] by *Dryfe* [222].

5th d. Black Bess [38] by Black Tom [37].

6th d. *Bell* [9] Imported.

BULLS.

[616] ROLLY-POLY.

Calved July 12th, 1882.

Bred by and the property of Mr. Geo. Coleman, Howell, Co. Livingston, Mich., U.S.

Sire Lord Malcolm [614].
Dam Chloe [584] by Herd Laddie [494].
2nd d. Pride of the Dairy [497] by Hardfortune 2nd [255].
3rd d. Margaret [285] by *Dryfe* [222].
4th d. Black Bess [38] by Black Tom [37].
5th d. *Bell* [9] Imported.

[617] COMET 2ND.

Calved November 7th, 1882.

Bred by Mr. Peter Davy, Ashippun, Co. Dodge, Wis., U.S.; the property of Mr. Samuel Martin, Eldora, Hardin, Iowa, U.S.

Sire Johnny Cope [265].
Dam Rose [306] by Sam [280].
2nd d. Sis [203] by Hardfortune [154].
3rd d. Susan [157] by *Dryfe* [222].
4th d. Bess [125] by Marquis [19].
5th d. *Chloe* [4] Imported.

[618] JOHNNY COPE 5TH.

Calved February 11th, 1882.

Bred by and the property of Mr. Philo Lasher, Coffeysburg, Co. Davies, Mo., U.S.

Sire Johnny Cope [265].
Dam Black Baby [451] by Johnny Cope [265].
2nd d. Blue Bell [438] by Duncan [437].
3rd d. Belle Mahone [196] by Our John [106].
4th d. Nellie Grey [109] by Black Jock [18].
5th d. Jessie [49] by Black Jock [18].
6th d. *Maggie* [48] by *Moss Trooper 2nd* [48] (151).

[619] SIR WILLIAM 2ND.

Calved June 9th, 1882.

Bred by and the property of **Mr. Philo Lasher, Coffeysburg, Co. Davies, Mo., U.S.**

Sire Sir William [340].
Dam Snowdrift [444] by Napier [346].
2nd d. Snowball [314] by Victor [122].
3rd d. Blooming Heather 2nd [80] by Dred [15].
4th d. *Blooming Heather* [53] by *Moss Trooper 2nd* [48] (151).
5th d. *Mary* by *Fergy* (19).
6th d. *Jane* by *Bob*.

[620] JOHN BORLAND.

Calved December 18th, 1876.

Bred by Mr. R. G. Hart, Lapeer, Michigan, U.S.; the property of Mr. Ashly Hamilton, Butler, Co. Bates, Mo., U.S.

Sire Shoo Fly [184].
Dam Dairymaid [286] by *Dryfe* [222].
2nd d. Black Bess [38] by Black Tom [37].
3rd d. *Bell* [9] Imported.

[621] JOHNNY SCOTT 3RD.

Calved April 4th, 1881.

Bred by Mr. Peter Davy, Monterey, Wis., U.S.; the property of Mr. Ashly Hamilton, Butler, Co. Bates, Mo., U.S.

Sire Johnny Scott [313].
Dam Snow [319] by Johnny Scott [313].
2nd d. Snowflake [315] by Johnny Scott [313].
3rd d. Snowball [314] by Victor [122].
4th d. Blooming Heather 2nd [80] by Dred [15].
5th d. *Blooming Heather* [53] by *Moss Trooper 2nd* [48] (151).
6th d. *Mary* by *Fergy* (19).
7th d. *Jane* by *Bob*.

BULLS.

[622] ### MOREAU.

Calved April 5th, 1882.

Bred by Mr. J. J. Rodgers, Abingdon, Ill., U.S.; the property of Mr. Ashly Hamilton, Butler, Co. Bates, Mo., U.S.

Sire John Borland [620].
Dam Snow [319] by Johnny Scott [313].
2nd d. Snowflake [315] by Johnny Scott [313].
3rd d. Snowball [314] by Victor [122].
4th d. Blooming Heather 2nd [80] by Dred [15].
5th d. *Blooming Heather* [53] by *Moss Trooper 2nd* [48] (151).
6th d. *Mary* by *Fergy* (19).
7th *Jane* by *Bob*.

[623] ### ALLAN.

Calved May 10th, 1882.

Bred by and the property of Mr. Geo. Whitfield, Rougemont, Que.

Sire *Scottish Borderer* (669).
Dam *Ella of Chapelhill* [627] (3669) by *Queensberry* (1029).
2nd d. *Baroness of Culmain* (2780) by *Mangerton* (525).
3rd d. *Miss Crocket*, in Scotland.

[624] ### CLANSMAN.

Calved April 10th, 1882.

Bred by and the property of Mr. Geo. Whitfield, Rougemont, Que.

Sire *Scottish Borderer* (669).
Dam *Violet of Chapelhill* [634] (3671) by *Queensberry* (1027).
2nd d. *Lizzie of Culmain* (2778) by *Clifton* (277).
3rd d. *Cherry* from the herd of Mr. John Cannon, Congeith, Kirkgunzeon, Dumfries, Scotland.

[625] ### KING CHARLES.

Calved January 20th, 1881.

Bred by and the property of Mr. Geo. Whitfield, Rougemont, Que.

Sire *Knowsley* (1279].
Dam *Meg Merrilees of Knocklae* [632] (2966) by *Neil Gow* (1138).
2nd d. *Lily of Knocklae* (2867) by *Blucher* (1136).
3rd d. *Old Lily* from the herd of Mr. Thomas Corrie, Knocklae, Balmaclellan, New Galloway, Scotland.

[626]   *LADDIE.*

Calved April, 1879.

Bred by the Earl of Galloway, Galloway House, Garlieston, Scotland; imported by and the property of Mr. Geo. Whitfield, Rougemont, Que.

Sire *Scottish Borderer* (669).
Dam a pure cow bred by Mr. John Cunningham, White Cairn, Scotland, and got by *Nelson* (1104).

[627]   *PREMIER OF LAIRDLAUGH (1629).*

Calved February 10th, 1880.

Bred by Messrs. J. S. & A. Nivison, Lairdlaugh, Dalbeattie, Scotland; imported by and the property of Mr. George Whitfield, Rougemont, Que.

Sire *Fancy Jock* (1177).
Dam *Primrose of Lairdlaugh* (3066) by *Tory of Lairdlaugh* (1057).
2nd d. *Daisy of Lairdlaugh* (2731) by *Border Knight of Pelutho* (625).
3rd d. *Beauty of Pelutho 2nd* (1635) by *Camerton* (542).
4th d. *Beauty of Pelutho 1st* (1358) by *Geordie 2nd of Green* (1135).

[628]   BURNS.

Calved March 20th, 1882.

Bred by Mr. Geo. Coleman, Howell, Co. Livingston, Mich, U.S.; the property of Mr. John J. Bush, Lansing, Mich., U.S.

Sire Lord Malcolm [614].
Dam Monah [641] by Rob Roy [501].
2nd d. Chloe [584] by Herd Laddie [494].
3rd d. Pride of the Dairy [497] by Hardfortune 2nd [255].
4th d. Margaret [285] by *Dryfe* [222].
5th d. Black Bess [38] by Black Tom [37].
6th d. *Bell* [9] Imported.

BULLS.

[629]   IONA.

Calved May 10th, 1881.

Bred by and the property of Mr. John J. Bush, Lansing, Mich., U.S.

Sire Blackbird [633].

Dam Lady Mitchell [649] by Robin Hood [349].
2nd d. Ingham Beauty [637] by Robin Hood [349].
3rd d. Maggie Lauder [110] by Black Jock [18].
4th d. Empress [50] by *Jock* [10].
5th d. *Beauty* [11] Imported.

[630]   MONTROSE.

Calved April 8th, 1882.

Bred by Mr. George Coleman, Howell, Co. Livingston, Mich., U.S.; the property of Mr. John J. Bush, Lansing, Mich., U.S.

Sire Lord Malcolm [614].
Dam Minnie [640] by Rob Roy [501].
2nd d. Maid of Livingston [616] by Napier [346].
3rd d. Pride of the Dairy [497] by Hardfortune 2nd [255].
4th d. Margaret [285] by *Dryfe* [222].
5th d. Black Bess [38] Black Tom [37].
6th d. *Bell* [9] Imported.

[631]   DUKE OF ARGYLE.

Calved May 30th, 1881.

Bred by Messrs. L. S. and W. B. Hall, Wakeman, Co. Huron, Ohio, U.S.; the property of Mr. Edmund Holdren, Grape Island, Co. Pleasants, W. Va., U.S.

Sire Lorne [447].
Dam Lady Princess [644] by Nelson [435].
2nd d. Gipsy Queen [194] by Our John [106].
3rd d. Dido [88] by Victor [122].
4th d. Magnolia [84] by Black Jock [18].
5th d. Galloway Lass [42] by *Jock* [10].
6th d. *Jet* [2] Imported.

[632] HURON LAD.

Calved October 11th, 1882.

Bred by and the property of Messrs. L. S. and W. B. Hall, Wakeman, Co. Huron, Ohio, U.S.

Sire Lorne [447].
Dam Lady Princess [644] by Nelson [435].
2nd d. Gipsy Queen [194] by Our John [106].
3rd d. Dido [88] by Victor [122].
4th d. Magnolia [84] by Black Jock [18].
5th d. Galloway Lass [42] by *Jock* [10].
6th d. *Jet* [2] Imported.

[633] BLACKBIRD.

Calved March 10th, 1877.

Bred by Mr. J. N. Smith, Bath, Co. Clinton, Mich., U.S.; the property of Mr. Saml. A. Browne, Pentwater, Co. Oceana, Mich., U.S.

Sire Robin Hood [349].
Dam Maggie Lauder 3rd [353] by Robin Hood [349].
2nd d. Maggie Lauder [110] by Black Jock [18].
3rd d. Empress [50] by *Jock* [10].
4th d. *Beauty* [11] Imported.

[634] GARFIELD.

Calved February 29th, 1882.

Bred by Mr. George Coleman, Howell, Co. Livingston, Mich., U.S.; the property of Mr. Saml. A. Browne, Pentwater, Co. Oceana, Mich., U.S.

Sire Lord Malcolm [614].
Dam Dinah [613] by Robin Hood [349].
2nd d. Chloe [584] by Herd Laddie [494].
3rd d. Pride of the Dairy [497] by Hardfortune 2nd [255].
4th d. Margaret [285] by *Dryfe* [222].
5th d. Black Bess [38] by Black Tom [37].
6th d. *Bell* [9] Imported.

BULLS.

[635] 2ND LORD MALCOLM, JR.

Calved June 29th, 1882.

Bred by Mr. Geo. Coleman, Howell, Co. Livingston, Mich., U.S.; the property of Mr. Chas. C. Wood, Beauro, Co. Lamoile, Ill., U.S.

Sire Lord Malcolm [614].
Dam Naomi [619] by Napier [346].
2nd d. Chloe [584] by Herd Laddie [494].
3rd d. Pride of the Dairy [447] by Hardfortune 2nd [255].
4th d. Margaret [285] by *Dryfe* [222].
5th d. Black Bess [38] by Black Tom [37].
6th d. *Bell* [9] by Imported.

[636] DAINTY DAVY.

Calved June 1st, 1881.

Bred by Mr. Thos. McCrae, Guelph, Co. Wellington, Ont.; the property of Messrs. J. W. Duncan & Sons, Independence, Co. Jackson, Mo., U.S.

Sire Lord Lochinvar [405].
Dam Lady Heron [180] by John A. [137].
2nd d. Minnie [178] by Dred [15].
3rd d. *Heather Bell* [12] Imported.

[637] RODERICK DHU.

Calved August 28th, 1882.

Bred by and the property of Mr. H. A. Disbrow, Atlantic, Co. Cass, Iowa, U.S.

Sire Rob Roy [501].
Dam Black Beauty [500] by Robin Hood [349].
2nd d. Queen of the Dairy [495] by Our John [106].
3rd d. Maggie Lauder [110] by Black Jock [18].
4th d. Empress [50] by *Jock* [10].
5th d. *Beauty* [11] Imported.

[638] BISMARCK.

Calved July 8th, 1874.

Bred by and the property of Mr. Wm. Hood, Guelph, Co. Wellington, Ont.

Sire Rodger [198].
Dam Gipsy [225] by Moss Trooper 4th [493].
2nd d. Beauty [224] by Prince Albert [190].
3rd d. Black Swan [219] by Prince Albert [190].
4th d. Black Jessie [218] by *Jock* [10].
5th d. *Sall* [14] Imported.

[639] CURLY JOCK.

Calved March 28th, 1878.

Bred by Mr. William Griffith, Little Current, Manitoulin; the property of Mr. F. McHardy, Emporia, Kansas, U.S.

Sire Bismarck [638].
Dam Gipsy [225] by Moss Trooper 4th [493].
2nd d. Beauty [224] by Prince Albert [190].
3rd d. Black Swan [219] by Prince Albert [190].
4th d. Black Jessie [218] by *Jock* [10].
5th d. *Sall* [14] Imported.

[640] MODOCK.

Calved May 27th, 1875.

Bred by Mr. R. G. Hart, Lapeer, Mich., U.S.; the property of Mr. Anderson De Witt, Washington Court House, Ohio, U.S.

Sire Shoo-Fly [184].
2nd d. Black Bess 2nd [284] by Johnny Cope [283].
3rd d. Margaret [285] by *Dryfe* [222].
4th d. Black Bess [38] by Black Tom [37].
5th d. *Bell* [9] Imported.

[641] BLACK DANE.

Calved July 10th, 1880.

Bred by Mr. Thomas McCrae, Guelph, Co. Wellington, Ont.; the property of Mr. Wm. Carter, San Belle Plains, Kansas, U.S.

Sire Major Gray [273].
Dam Dagmar [153] by Black Jock [18].
2nd d. Jennie [152] by Garibaldi [130].
3rd d. Victoria [150] by *Black Prince* [152].
4th d. *Beauty* [151] by *Peter* (116).

[642] LA FAYETTE PRINCE.

Calved April 4th, 1882.

Bred by and the property of Messrs. Isaac B. Lutz & Son, La Fayette, Ind., U.S.

Sire Bob Wooley [410].
Dam Dora 2nd [293] by Shoo-Fly [184].
2nd d. Dora [288] by Shoo-Fly [184].
3rd d. Dairymaid [286] by *Dryfe* [222].
4th d. Black Bess [38] by Black Tom [37].
5th d. *Bell* [9] Imported.

COWS.

[1] ### *BONNY.*
Calved in 1851.

Imported in 1853 by Mr. Graham, Vaughan, Co. York, Ont., from the herd of Mr. John Carruthers, Kirkhill, Dumfries, Scotland.

Produce: 1854, Norval [72]; 1855, Black Tom [37]; 1857, Bonny 2nd [36]; 1858, Venus [73]; 1862, Sall [170].

[2] ### *JET.*
Calved in 1851.

Imported in 1853 by Mr. Graham, Vaughan, Co. York, Ont., from the herd of Mr. Graham, Shaw Dryfe, Dumfries, Scotland.

Produce: 1856, Galloway Lass [42]; 1857, Jock 2nd [45].

[3] ### *VENUS.*
Calved in 1851.

Imported in 1853 by Mr. Graham, Vaughan, Co. York, Ont., from the herd of Mr. Graham, Shaw Dryfe, Dumfries, Scotland.

Produce: 1860, Nelson [47]; 1861, Curly Bob [44]; 1862, Maggie [64].

[4] ### *CHLOE.*
Calved in 1851.

Imported in 1853 by Mr. Graham, Vaughan, Co. York, Ont., from the herd of Mr. Robert Brown, Capaylann, Dumfries, Scotland.

Produce: 1857, Bonny [74]; 1859, Bess [125]; 1861, Molly [17]; 1862, McQuhon [26]; 1864, Coquette [119]; 1865, Lady Isabella [100]; 1867, Maggie Lauder [148]; 1870, Topsy [200].

[5] *WHITE BAG.*

Calved in 1851.

The property of Mr. John Fleming, Vaughan, Co. York, Ont.; imported in 1853 by Mr. Graham, Vaughan, Co. York, Ont., from the herd of Mr. Rogerson, Leighton Hall, Scotland.

Produce: 1854, Black Jock [18]; 1855, Bonny [27]; 1856, William Wallace *alias* Clover Hill [28]; 1857, Grampion [29]; 1858, Susan [30]; 1859, Lucy [31]; 1862, Countess [34]; 1865, Jenny Lind [32].

[6] *BLACK BESS.*

Calved in 1851.

The property of Mr. Torrence, Vaughan, Co. York, Ont.; imported in 1853 by Mr. Graham, Vaughan, Co. York, Ont., from the herd of Mr. Rogerson, Leighton Hall, Dumfries, Scotland.

Produce: 1858, Queen of Beauty [95]; 1865, Jet [343]; 1866, Sall [192]; Black Jane [210].

[7] *PHILLIS.*

Calved in 1852.

The property of Mr. J. Jardine, Saltfleet, Co. Wentworth, Ont.; imported in 1853 by Mr. Graham, Vaughan, Co. York, Ont., from the herd of Mr. Carruthers, Kirkhill, Dumfries, Scotland.

[8] *BLACKY.*

Calved in 1852.

The property of Mr. Geo. Miller, Markham, Co. York, Ont.; imported in 1853 by Mr. Graham, Vaughan, Co. York, Ont., from the herd of Mr. Carruthers, Kirkhill, Dumfries, Scotland.

Produce: 1861, Mrs. Snell [105]; 1863, Ebony [177].

[9] *BELL.*

Calved in 1852.

The property of Mr. Arthur McNeil, Vaughan, Co. York, Ont.; imported in 1853 by Mr. Graham, Vaughan, Co. York, Ont., from the herd of Mr. Carruthers, Kirkhill, Dumfries, Scotland.

Produce: 1859, Black Bess [38]; 1861, Young Bonny [39]; 1862, Honest Tom [98]; 1863, Young Blucher [41]; 1864, Wonderful Lad [214]; 1867, Chub [115].

COWS.

[11] *BEAUTY.*

Calved in 1854.

Imported in 1856 by Mr. Graham, Vaughan, Co. York, Ont., from the herd of Mr. Carruthers, Kirkhill, Dumfries, Scotland.

Produce: 1857, Moss Trooper [20]; 1858, Empress [50]; 1859, Prince of Wales [57]; 1861, Lady Mary [16]; 1862, Lady Marion [186].

[12] *HEATHER BELL.*

Calved in 1854.

The property of Mr. Thos. McCrae, Guelph, Co. Wellington, Ont.; imported in 1856 by Mr. Graham, Vaughan, Co. York, Ont., from the herd of Mr. Carruthers, Kirkhill, Dumfries, Scotland.

Produce: 1860, Heather Bell 2nd [25]; 1861, Tam O'Shanter [24]; 1862, Lord Byron [23]; 1863, Clara [126]; 1866, Minnie [178].

[13] *TOPSY.*

Calved in 1854.

Imported in 1856 by Mr. Graham, Vaughan, Co. York, Ont., from the herd of Mr. Graham, Shaw Dryfe, Dumfries, Scotland.

Produce: 1857, Marquis [19]; 1858, Victoria [135]; 1859, Prince Albert [190].

[14] *SALL.*

Calved in 1854.

Imported in 1856 by Mr. Graham, Vaughan, Co. York, Ont., from the herd of Mr. Graham, Shaw Dryfe, Dumfries, Scotland.

Produce: 1857, Shaw [21]; 1861, Dred [15]; 1862, Duncan [64]; 1863, Sall 2nd [79]; 1864, Nancy [76]; 1865, Our John [106]; 1866, Robin Hood [144]; 1867, Uncle Tom [107]; 1869, Robin [166].

[16] LADY MARY.

Calved in 1861.

Bred by and the property of Mr. J. Graham, Vaughan, Co. York, Ont

Sire Black Jock [18].

Dam *Beauty* [11] Imported.

[17] MOLLY.

Calved January, 1861.

Bred by Mr. J. Graham, Vaughan, Co. York, Ont.; the property of Mr. Thos. McCrae, Guelph, Co. Wellington, Ont.

Sire Black Jock [18].
Dam *Chloe* [4] Imported.

Produce : 1865, Kate [176]; 1866, Em [144]; 1867, Nellie [160].

[21] SHAW.

Calved in 1875.

Bred by Mr. Graham, Vaughan, Co. York, Ont.; the property of Mr. Alex. Kerr, Westminster, Co. Middlesex, Ont.

Sire *The Pilgrim* [32] (32).
Dam *Sall* [14] Imported.

Produce : 1861, Flora [22]; 1867, Rose [132].

[22] FLORA.

Calved October, 1861.

Bred by Mr. Norman McLeod, King, Co. York, Ont.; the property of Mr. J. Graham, Vaughan, Co. York, Ont.

Sire Black Jock [18].
Dam Shaw [21] by *The Pilgrim* [32] (32).
2nd d. *Sall* [14] Imported.

Produce : 1864, Louisa [86]; 1866, Victor Hugo [92].

[25] HEATHER BELL 2ND.

Calved May 4th, 1860.

Bred by Mr. James Graham, Vaughan, Co. York, Ont.; the property of Mr. John Dunlop, Woodstock, Co. Oxford, Ont.

Sire Black Tom [37].
Dam *Heather Bell* [12] Imported.

Produce : 1863, Sambo [71]; 1864, Kenmure [112].

COWS.

[27]   BONNY.

Calved March, 1855.

Bred by Mr. John Fleming, Vaughan, Co. York, Ont. ; the property of Mr. John Snell, Chinguacousy, Co. Peel, Ont.

Sire *Jock* [10].

Dam *White Bag* [5] Imported.

Produce : 1860, Fred Douglas [46]; 1861, Pocahontas [60]; 1862, Robert Bruce [66]; 1863, St. Clair [120]; 1864, Tecumseh [117]; 1865, Dairymaid [158]; 1866, Selkirk [101]; 1867, Bonny Dundee [136]; 1868, Sir William Wallace [138].

[30]   SUSAN.

Calved May, 1858.

Bred by and the property of Mr. J. Fleming, Vaughan, Co. York, Ont.

Sire Black Jock [18].

Dam *White Bag* [5] Imported.

Produce : 1862, Helen Douglas [33]; 1864, Young Dryfe [127]; 1865, Cariboo [90]; 1866, El Hakim [91].

[31]   LUCY.

Calved August 3rd, 1859.

Bred by Mr. John Fleming, Vaughan, Co. York, Ont. ; the property of Mr. John Dunlop, Woodstock, Co. Oxford, Ont.

Sire Black Jock [18].

Dam *White Bag* [5] Imported.

[32]   JENNY LIND.

Calved November, 1865.

Bred by and the property of Mr. John Fleming, Vaughan, Co. York, Ont.

Sire Marquis [19].

Dam *White Bag* [5] Imported.

Produce : 1868, Robin Hood [156].

[33]   HELEN DOUGLAS.

Calved May, 1862.

Bred by and the property of Mr. John Fleming, Vaughan, Co. York, Ont.

Sire Marquis [19].

Dam Susan [30] by Black Jock [18].

2nd d. *White Bag* [5] Imported.

[34] COUNTESS.

Calved November, 1862.

Bred by and the property of Mr. John Fleming, Vaughan, Co. York, Ont.
Sire Marquis [19].
Dam *White Bag* [5] Imported.

[35] MARY.

Calved June, 1862.

Bred by and the property of Mr. John Fleming, Vaughan, Co. York, Ont.
Sire Marquis [19].
Dam Queen of Beauty [95] by Black Jock [18].
2nd d. *Black Bess* [6] Imported.

Produce: 1867, Hyena [185].

[36] BONNY 2ND.

Calved April 15th, 1857.

Bred by Mr. John Fleming, Vaughan, Co. York, Ont.; the property of Mr. Thos. Chaffey, Acton, Co. Halton, Ont.
Sire Black Jock [18].
Dam *Bonny* [1] Imported.

Produce: 1861, Uncle Tom [58].

[38] BLACK BESS.

Calved April, 1859.

Bred by and the property of Mr. Arthur McNeil, Vaughan, Co. York, Ont.
Sire Black Tom [37].
Dam *Bell* [9] Imported.

Produce: 1863, Woolwich Queen [96]; 1867, Jock 2nd [207]; 1868, Dairymaid [286]; 1869, Margaret [285].

[39] YOUNG BONNY.

Calved March, 1861.

Bred by and the property of Mr. Arthur McNeil, Vaughan, Co. York, Ont.
Sire Black Tom [37].
Dam *Bell* [9] Imported.

Produce: 1862, Young Matchless [40]; 1868, Duncan [437]; 1869, Napier [346]; 1871, Beauty [258].

COWS.

[42]   GALLOWAY LASS.

Calved in 1856.

Bred by Mr. Graham, Vaughan, Co. York, Ont.; the property of Mr. John Torrence.

Sire *Jock* [10].
Dam *Jet* [2] Imported.

Produce: 1858, Dairymaid [51]; 1859, Lucy [52]; 1860, Blooming Beauty [54]; 1861, Queen of the West [43]; 1863, Magnolia [84].

[43]   QUEEN OF THE WEST.

Calved June, 1861.

Bred by and the property of Mr. John Torrence, Vaughan, Co. York, Ont.

Sire Black Jock [18].
Dam Galloway Lass [42] by *Jock* [10].

2nd d. *Jet* [2] Imported.

[48]   *MAGGIE.*

Calved in 1854.

Imported in 1855 by Mr. Wm. Miller, Pickering, Co. Ontario, Ont., from the herd of Mr. Jas. Beattie, Newbie House, Dumfries, Scotland.

Sire *Moss Trooper 2nd* [48] (151).

Produce: 1857, Jessie [49]; 1859, Jessie Miller [98]; 1863, Eva [43]; 1874, Pride of the Dairy [497].

[49]   JESSIE.

Calved April 9th, 1857.

Bred by Mr. Wm. Miller, Pickering, Co. Ontario, Ont.; the property of Mr. John Snell, Chinguacousy, Co. Peel, Ont., afterwards of Mr. Thos. McCrae, Guelph, Co. Wellington, Ont.

Sire Black Jock [18].
Dam *Maggie* [48] by *Moss Trooper 2nd* [48] (151).

Produce: 1862, Cherry [81]; 1863, Roderick Dhu [68]; 1864, Idaho [97]; 1865, Grace [99]; 1866, Nelly Grey [109]; 1867, Grace Darling [145]; 1868, Cooderina [161].

[50]   EMPRESS.

Calved March 6th, 1858.

Bred by Mr. John Torrence, Vaughan, Co. York, Ont.; the property of Mr. John Snell, Chinguacousy, Co. Peel, Ont.

Sire *Jock* [10].

Dam *Beauty* [11] Imported.

Produce: 1861, Lavina [55]; 1862, Malcolm [62]; 1863, Annie Laurie [82]; 1864, William Wallace [67]; 1865, Louan [87]; 1866, Gipsy Queen [89]; 1867, Maggie Lauder [110]; 1868, Mary [111]; 1869, Queen of the West [168]; 1870, Prince Albert [239]; 1871, May Queen [237].

[51]   DAIRYMAID.

Calved March 1st, 1858.

Bred by Mr. John Torrence, Vaughan, Co. York, Ont.; the property of Mr. John Snell, Chinguacousy, Co. Peel, Ont.

Sire Black Jock [18].

Dam Galloway Lass [42] by *Jock* [10].

2nd d. *Jet* [2] Imported.

Produce: 1861, Beauty [59]; 1862, Mary Grey [65]; 1863, Monitor [121]; 1864, El Dorado [118].

[52]   LUCY.

Calved April 15th, 1859.

Bred by and the property of Mr. John Torrence, Vaughan, Co. York, Ont.

Sire Black Jock [18].

Dam Galloway Lass [42] by *Jock* [10].

2nd d. *Jet* [2] Imported.

Produce: 1861, Topsy 2nd [56]; 1862, Ruby [61].

[53]   *BLOOMING HEATHER.*

Calved March 26th, 1860.

Imported by and the property of Mr. Simon Beattie, Markham, Co. York, Ont., from the herd of Mr. James Beattie, Newbie House, Annan, Scotland.

Sire *Moss Trooper 2nd* [48] (151).

Dam *Mary* by *Fergy* (19).

2nd d. *Jane* by *Bob*.

Produce: 1861, Young Scotland [63]; 1863, Blooming Heather 2nd [80]; 1864, Heather Bell 3rd [85].

COWS.

[54]   BLOOMING BEAUTY.

Calved March 3rd, 1860.

Bred by Mr. John Torrence, Vaughan, Co. York, Ont.; the property of Mr. John Snell, Chinguacousy, Co. Peel, Ont.
Sire Black Jock [18].
Dam Galloway Lass [42] by *Jock* [10].
2nd d. *Jet* [2] Imported.

Produce: 1864, Victor [122].

[55]   LAVINA.

Calved 6th March, 1861.

Bred by Mr. John Torrence, Vaughan, Co. York, Ont.; the property of Mr. John Snell, Chinguacousy, Co. Peel, Ont.
Sire Black Jock [18].
Dam Empress [50] by *Jock* [10].
2nd d. *Beauty* [11] Imported.

Produce: 1864, Donald [123]; 1865, Cricket [209]; 1866, Tonawanda [93]; 1869, Nina [187].

[56]   TOPSY 2ND.

Calved April 15th, 1861.

Bred by Mr. John Snell, Chinguacousy, Co. Peel, Ont.; the property of Mr. Robert Collacutt, Darlington, Co. Durham, Ont.
Sire Prince of Wales [57].
Dam Lucy [52]   Black Jock [18].
2nd d. Galloway Lass [42] by *Jock* [10].
3rd d. *Jet* [2] Imported.

[59]   BEAUTY.

Calved May 15th, 1861.

Bred by Mr. Henry J. Boulton, Humberford, Co. York, Ont.; the property of Mr. Geo. Anderson, Varna, Co. Huron, Ont.
Sire Prince of Wales [57].
Dam Dairymaid [51] by Black Jock [18].
2nd d. Galloway Lass [42] by *Jock* [10].
3rd d. *Jet* [2] Imported.

[60] POCAHONTAS.

Calved December 29th, 1861.

Bred by Mr. John Snell, Chinguacousy, Co. Peel, Ont.; the property of Mr. Thos. McCrae, Guelph, Co. Wellington, Ont.

Sire Marquis [19].
Dam Bonny [27] by Black Jock [18].
2nd d. *White Bag* [5] Imported.

Produce: 1865, Lady Kenmure [140]; 1866, Flora [141]; 1867, Black Douglas [148].

[61] RUBY.

Calved March 17th, 1862.

Bred by and the property of Mr. John Snell, Edmonton, Co. Peel, Ont.

Sire Fred Douglas [46].
Dam Lucy [52] by Black Jock [18].
2nd d. Galloway Lass [42] by *Jock* [10].
3rd d. *Jet* [2] Imported.

Produce: 1863, Elvira [83].

[64] MAGGIE.

Calved March, 1862.

Bred by Mr. R. Conway, Co. York, Ont.; the property of Mr. David Messenger, Cooksville, Co. York, Ont.

Sire Black Jock [18].
Dam *Venus* [3] Imported.

[65] MARY GREY.

Calved December 5th, 1862.

Bred by and the property of Mr. John Snell, Edmonton, Co. Peel, Ont.

Sire Fred Douglas [46].
Dam Dairymaid [51] by Black Jock [18].
2nd d. Galloway Lass [42] by *Jock* [10].
3rd d. *Jet* [2] Imported.

COWS.

[69] *DANDY.*

Calved 7th January, 1854.

Imported by and the property of Mr. George Roddick, Cobourg, Co. Northumberland, Ont., from the herd of Messrs. W. & J. Shannon, Balig, Scotland.

Sire *The Earl* [69].
Dam *Black Jess* [6] by *Geordie of Riggfoot* (234).
2nd d. *Blacky* [8] by *Galloway Tam* (100).

Produce: 1858, Morrach [78].

[73] VENUS.

Calved in 1858.

Bred by Mr. Graham, Vaughan, Co. York, Ont.; the property of Mr. Robert Conway, Co. York, Ont.

Sire *Jock* [10].
Dam *Bonny* [1] Imported.

Produce: 1864, Brenda [77].

[74] BONNY 2ND.

Calved in 1857.

Bred by Mr. Graham, Vaughan, Co. York, Ont.; the property of Mr. Robert Conway, Co. York, Ont.

Sire *Jock* [10].
Dam *Chloe* [4] Imported.

Produce: 1864, Hardfortune [154]; 1866, Lizzie [114]; 1868, Prince [155].

[75] *NEWBIE LASS.*

Calved in 1856.

Owned by Mr. Thomas McCrae, Guelph, Co. Wellington, Ont., and imported by him from the herd of Mr. James Beattie, Newbie House, Annan, Scotland.

Sire *Moss Trooper* [74] (296).
Dam a pure Galloway cow in Scotland.

Produce: 1863, Maumee [92]; 1867, Duckystone [143]; 1870, Lord Kenmure [271].

[76]   NANCY.
Calved in 1864.
Bred by Mr. R. Conway, Co. York, Ont.; the property of Mr. Michael Murphy, Enniskillen, Co. Lambton, Ont.
Sire Norval [72].
Dam *Sall* [14] Imported.

[77]   BRENDA.
Calved in 1864.
Bred by Mr. Robt. Conway, Co. York, Ont.; the property of Mr. M. Murphy, Enniskillen, Co. Lambton, Ont.
Sire Norval [72].
Dam Venus [73] by *Jock* [10].
2nd d. *Bonny* [1] Imported.

[79]   SALL 2ND.
Calved February 13th, 1863.
Bred by and the property of the late Mr. John Snell, Edmonton, Co. Peel, Ont.
Sire Fred Douglas [46].
Dam *Sall* [14] Imported.

[80]   BLOOMING HEATHER 2ND.
Calved December 11th, 1863.
Bred by the late John Snell, Edmonton, Co. Peel, Ont.; the property of the Michigan State Agricultural College, Lansing, Mich., U.S.
Sire Dred [15].
Dam *Blooming Heather* [53] by *Moss Trooper 2nd* [48] (152).
2nd d. *Mary* by *Fergy* (19).
3rd d. *Jane* by *Bob*.
Produce: 1870, Snowball [314].

[81]   CHERRY.
Calved March 2nd, 1862.
Bred by and the property of the late John Snell, Edmonton, Co. Peel, Ont.
Sire Fred Douglas [46].
Dam Jessie [49] by Black Jock [18].
2nd d. *Maggie* [48] by *Moss Trooper 2nd* [48] (151).

COWS.

[82]   ANNIE LAURIE.

Calved November 26th, 1863.

Bred by and the property of the late John Snell, Edmonton, Co. Peel, Ont.

Sire Black Jock [18].
Dam Empress [50] by *Jock* [10].
2nd d. *Beauty* [11] Imported.

Produce: 1866, Saginaw [94]; 1867, Bessie Bell [146].

[83]   ELVIRA.

Calved November 16th, 1863.

Bred by and the property of the late John Snell, Edmonton, Co. Peel, Ont.

Sire Black Jock [18].
Dam Ruby [61] by Fred Douglas [46].
2nd d. Lucy [52] by Black Jock [18].
3rd d. Galloway Lass [42] by *Jock* [10].
4th d. *Jet* [2] Imported.

[84]   MAGNOLIA.

Calved May 21st, 1863.

Bred by and the property of the late John Snell, Edmonton, Co. Peel, Ont.

Sire Black Jock [18].
Dam Galloway Lass [42] by *Jock* [10].
2nd d. *Jet* [2] Imported.

Produce: 1865, Dido [88]; 1870, Magnolia 3rd [401].

[85]   HEATHER BELL 3RD.

Calved December 18th, 1864.

Bred by and the property of the late John Snell, Edmonton, Co. Peel, Ont.

Sire Black Jock [18].
Dam *Blooming Heather* [53] by *Moss Trooper 2nd* [48] (151).
2nd d. *Mary* by *Fergy* (19).
3rd d. *Jane* by *Bob*.

[86] LOUISA.

Calved November 6th, 1864.

Bred by Mr. Jas. Graham, Vaughan, Co. York, Ont.; the property of the late John Snell, Edmonton, Co. Peel, Ont.

Sire Dred [15].
Dam Flora [22] by Black Jock [18].
2nd d. Shaw [21] by *Pilgrim* (32).
3rd d. *Sall* [14] Imported.

[87] LOUAN.

Calved October 26th, 1865.

Bred by and the property of the late John Snell, Edmonton, Co. Peel, Ont.

Sire Black Jock [18].
Dam Empress [50] by *Jock* [10].
2nd *Beauty* [11] Imported.

Produce: 1868, Aikendrum [133].

[88] DIDO.

Calved April 10th, 1865.

Bred by and the property of the late John Snell, Edmonton, Co. Peel, Ont.

Sire Victor [122].
Dam Magnolia [84] by Black Jock [18].
2nd d. Galloway Lass [42] by *Jock* [10].
3rd d. *Jet* [2] Imported.

Produce: 1870, Gipsy Queen 2nd [194].

[89] GIPSY QUEEN.

Calved October 12th, 1866.

Bred by and the property of the late John Snell, Edmonton, Co. Peel, Ont.

Sire Black Jock [18].
Dam Empress [50] by *Jock* [10].
2nd d. *Beauty* [11] Imported.

COWS.

[92] MAUMEE.

Calved in 1863.

Bred by Mr. Geo. Miller, Markham, Co. York, Ont.; the property of Mr. Thomas McCrae, Guelph, Co. Wellington, Ont.

Sire *Jock* [10].
Dam *Newbie Lass* [75] Imported.

Produce: 1867, Little John [168]; 1870, Galloway Lass [234].

[93] EVA.

Calved September 28th, 1863.

Bred by Mr. Geo. Miller, Markham, Co. York, Ont.; the property of Mr. Thos. McCrae, Guelph, Co. Wellington, Ont.

Sire *Jock* [10].

Dam *Maggie* [48] by *Moss Trooper 2nd* [48] (151).

[95] QUEEN OF BEAUTY.

Calved June 15th, 1858.

Bred by Mr. John Torrence, Vaughan, Co. York, Ont.; the property of Mr. Thomas McCrae, Guelph, Co. Wellington, Ont.

Sire Black Jock [18].
Dam *Black Bess* [6] Imported.

Produce: 1862, Mary [35]; 1864, Blue Bell [235]; 1866, May Queen [143]; 1867, Sir John A. [137]; 1868, Black Prince [160].

[96] WOOLWICH QUEEN.

Calved October 7th, 1863.

Bred by Mr. W. H. Peterson, Guelph, Co. Wellington, Ont.; the property of Mr. Thomas McCrae, Guelph, Co. Wellington, Ont.

Sire William Wallace *alias* Clover Hill [28].
Dam Black Bess [38] by Black Tom [37].
2nd d. *Bell* [9] Imported.

Produce: 1867, Libbie [208]; 1868, Prince Lee Boo [232]; 1870, Roger [198]; 1871, Fancy [202].

[97] IDAHO.

Calved December 10th, 1864.

Bred by the late John Snell, Edmonton, Co. Peel, Ont.; the property of Mr. Thos. McCrae, Guelph, Co. Wellington, Ont.

Sire *Jock* [10].
Dam Jessie [49] by Black Jock [18].
2nd d. *Maggie* [48] by *Moss Trooper 2nd* [48] (151).
Produce: 1866, Cora [142]; 1870, Topsy Wopsy [189].

[98] JESSIE MILLER.

Calved in 1859.

Bred by Mr. Geo. Miller, Markham, Co. York, Ont.; the property of Mr. Thomas McCrae, Guelph, Co. Wellington, Ont.

Sire *Jock* [10].
Dam *Maggie* [48] by *Moss Trooper 2nd* [48] (151).
Produce: 1867, Bella [149].

[99] GRACE.

Calved November 25th, 1865.

Bred by and the property of Mr. Thos. McCrae, Guelph, Co. Wellington, Ont.

Sire Donald [123].
Dam Jessie [49] by Black Jock [18].
2nd d. *Maggie* [48] by *Moss Trooper 2nd* [48] (151).

[100] LADY ISABELLA.

Calved December 25th, 1865.

Bred by and the property of Mr. Thos. McCrae, Guelph, Co. Wellington, Ont.

Sire Donald [123].
Dam *Chloe* [4] Imported.
Produce: 1869, Johnny Cope [283]; 1870, Queen of Beauty [197]; 1872, Queen Vic. [236]; 1873, Hardfortune 2nd [255]; 1874, Sall [485].

[103] *HELEN.*

Imported by Mr. Geo. Miller, Markham, Co. York, Ont., from Scotland.
Produce: 1862, Salina [104].

COWS.

[104] SALINA.

Calved May 25th, 1862.

Bred by Col. Thompson, Aikenshaw, Co. York, Ont.; the property of Col. R. L. Denison, Dover Court, Toronto, Co. York, Ont.

Sire Prince of Wales [57].

Dam *Helen* [103] Imported.

Produce: 1865, Dick Peterson [219]; 1866, Queen's Own of Ridgeway [217]; 1867, John A. [102].

[105] MRS. SNELL.

Calved June 30th, 1861.

Bred by the late John Snell, Edmonton, Co. Peel, Ont.; the property of Col. R. L. Denison, Dover Court, Toronto, Co. York, Ont.

Sire Black Jock [18].

Dam *Blacky* [8] Imported.

Produce: 1867, Lord Monck [218].

[109] NELLY GREY.

Calved February 19th, 1866.

Bred by the late John Snell, Edmonton, Co. Peel, Ont.; the property of Mr. Wm. Hood, Guelph, Co. Wellington, Ont.

Sire Black Jock [18].

Dam Jessie [49] by Black Jock [18].

2nd d. *Maggie* [48] by *Moss Trooper 2nd* [48] (151).

Produce: 1868, Geordie [108]; 1869, Johnny Cope [165]; 1870, Shoo-Fly [184]; 1871, Belle Mahone [196]; 1873, Black Jock [253].

[110] MAGGIE LAUDER.

Calved January 8th, 1867.

Bred by the late John Snell, Edmonton, Co. Peel, Ont.; the property of Mr. William Hood, Guelph, Co. Wellington, Ont.

Sire Black Jock [18].

Dam Empress [50] by *Jock* [10].

2nd d. *Beauty* [11] Imported.

Produce: 1869, Rosy Hood [188]; 1870, Maggie Lauder 2nd [347]; 1871, Johnny Scott [313]; 1872, Queen of the Dairy [495]; 1873, Billy Hood [351]; 1875, Maggie Lauder 3rd [353]; 1876, Ingham Beauty [637]; 1877, Maggie 4th [356]; 1878, Pet [498].

[111]   MARY.

Calved July 6th, 1868.

Bred by and the property of Mr. Wm. Hood, Guelph, Co. Wellington, Ont.

Our John [106].
Dam Empress [50] by *Jock* [10].
2nd d. *Beauty* [11] Imported.

Produce: 1871, Adah [191]; 1876, Centennial [335]; 1877, Polly [514]; 1878, Little Mary [400]; 1879, Royal Charlie [474]; 1882, Prince Charlie [484].

[112]   KENMURE.

Calved April 7th, 1864.

Bred by and the property of Mr. John Dunlop, Woodstock, Co. Oxford, Ont.

Sire Morrach [78].
Dam Heather Bell 2nd [25] by Black Tom [37].
2nd d. *Heather Bell* [12] Imported.

[113]   *BARBARA*.

Imported by Mr. Wm. Roddick, Cobourg, Co. Northumberland, Ont., from the herd of Messrs. W. & J. Sherman, Balig, Scotland.

Produce: 1858, Uncle Tom [113].

[114]   LIZZIE.

Calved March 15th, 1866.

Bred by and the property of Mr. Arthur McNeil, Vaughan, Co. York, Ont.

Sire Dred [15].
Dam Bonny 2nd [74] by *Jock* [10].
2nd d. *Chloe* [4] Imported.

Produce: 1868, Margaret [344]; 1869, Rosa McNeil [345]; 1871, Fred [200].

[116]   AGNES 2ND.

Calved April 8th, 1866.

Bred by and the property of Mr. David Messenger, Cooksville, Co. Peel, Ont.

Sire Black Jock [18].
Dam Agnes [124] by Norval [72].
2nd d. Bess [125] by Marquis [19].
Dam *Chloe* [4] Imported.

COWS.

[119]   COQUETTE.

Calved November 13th, 1864.

Bred by Mr. James Graham, Vaughan, Co. York, Ont. ; the property of Mr. Thos. McCrae, Guelph, Co. Wellington, Ont.

Sire Dred [15].
Dam *Chloe* [4] Imported.

[120]   MAGGIE.

Calved June, 1865.

Bred by and the property of Mr. Thos. McCrae, Guelph, Co. Wellington, Ont.

Sire Dred [15].
Dam *Polly Shaw* [174] Imported.

[124]   AGNES.

Calved May, 1862.

Bred by Mr. Graham, Vaughan, Co. York, Ont.; the property of Mr. Alex. Mounsey, Etobicoke, Co. York, Ont.

Sire Norval [72].
Dam Bess [125] by Marquis [19].
2nd d. *Chloe* [4] Imported.

Produce : 1865, Garibaldi [130] ; 1866, Agnes 2nd [116] ; 1870, Jean [201].

[125]   BESS.

Calved July 28th, 1859.

Bred by Mr. J. Graham, Vaughan, Co. York, Ont. ; the property of Mr. Alex. Mounsey, Etobicoke, Co. York, Ont.

Sire Marquis [19].
Dam *Chloe* [4] Imported.

Produce : 1862, Agnes [124] ; 1863, Elsie [128] ; 1865, Fanny [129] ; 1868, Susan [157].

[126]   CLARA.

Calved July, 1863.

Bred by Mr. Graham, Vaughan, Co. York, Ont. ; the property of Mr. Alex. Mounsey, Etobicoke, Co. York, Ont.

Sire Black Jock [18].
Dam *Heather Bell* [12] Imported.

Produce : 1865, Hannah [131].

[128] ELSIE.

Calved July 28th, 1863.

Bred by and the property of Mr. Alex. Mounsey, Etobicoke, Co. York, Ont.

Sire Black Jock [18].
Dam Bess [125] by Marquis [19].
2nd d. *Chloe* [4] Imported.

[129] FANNY.

Calved January 28th, 1865.

Bred by and the property of Mr. Alex. Mounsey, Etobicoke, Co. York, Ont.

Sire Dred [15].
Dam Bess [125] by Marquis [19].
2nd d. *Chloe* [4] Imported.

[131] HANNAH.

Calved May 10th, 1865.

Bred by and the property of Mr. Alex. Mounsey, Etobicoke, Co. York, Ont.

Sire Dred [15].
Dam Clara [126] by Black Jock [18].
2nd d. *Heather Bell* [12] Imported.

[132] ROSE.

Calved in 1867.

Bred by and the property of Mr. John Wilson, Westminster, Co. Middlesex, Ont.

Sire Young Blucher [41].
Dam Shaw [21] by *The Pilgrim* [32].
2nd d. *Sall* [14] Imported.

[135] VICTORIA.

Calved May 2nd, 1858.

Bred by and the property of Mr. Joseph Jardine, Saltfleet, Co. Wentworth, Ont.

Sire *Jock* [10].
Dam *Topsy* [13] Imported.

COWS.

[136] *BONNY.*

Bred by and the property of Mr. W. Pagan, Duncrief, Scotland.
Produce: 1859, Saltfleet [134].

[137] HEATHER BELL 4TH.

Calved November, 1864.

Bred by Mr. Geo. Miller, Markham, Co. York, Ont.; the property of Mr. Thomas McCrae, Guelph, Co. Wellington, Ont.

Sire Young Scotland [63].
Dam *Polly Shaw* [174] Imported.

[140] LADY KENMURE.

Calved February 16th, 1865.

Bred by and the property of Mr. Thomas McCrae, Guelph, Co. Wellington, Ont.

Sire *Jock* [10].
Dam Pocahontas [60] by Marquis [19].
2nd d. Bonny [27] by *Jock* [10].
3rd d. *White Bag* [5] Imported.

Produce: 1878, Jock O'Bombie [139]; 1870, Lady Gordon [182]; 1871, Pocahontas [166]; 1872, Lady Lisgar [241]; 1873, Dufferin [266]; 1874, Lady Hamilton [274].

[141] FLORA.

Calved January 23rd, 1866.

Bred by and the property of Mr. Thomas McCrae, Guelph, Co. Wellington, Ont.

Sire Donald [123].
Dam Pocahontas [60] by Marquis [19].
2nd d. Bonny [27] by *Jock* [10].
3rd d. *White Bag* [5] Imported.

[142] CORA.

Calved June 12th, 1866.

Bred by and the property of Mr. Thomas McCrae, Guelph, Co. Wellington, Ont.

Sire Saginaw [94].
Dam Idaho [97] by Black Jock [18].
2nd d. Jessie [49] by *Jock* [10].
3rd d. *Maggie* [48] Imported.

[143] MAY QUEEN.

Calved April 28th, 1866.

Bred by Mr. Thos. McCrae, Guelph, Co. Wellington, Ont.; the property of Mr. Wm. Dow, Nichol, Co. Wellington, Ont.

Sire Honest Tom [98].
Dam Queen of Beauty [95] by Black Jock [18].
Dam *Black Bess* [6] Imported.

Produce: 1870, King Tom [175]; 1871, Queen Mab [206]; 1872, Annie Laurie [251]; 1875, Sleeping Maggie [433].

[144] EM.

Calved November 20th, 1866.

Bred by and the property of Mr. Thomas McCrae, Guelph, Co. Wellington, Ont.

Sire Dred [15].
Dam Molly [17] by Black Jock [18].
2nd d. *Chloe* [4] Imported.

[145] GRACE DARLING.

Calved October 28th, 1866.

Bred by and the property of Mr. Thos. McCrae, Guelph, Co. Wellington, Ont.

Sire Dred [15].
Dam Jessie [49] by Black Jock [18].
2nd d. *Maggie* [48] by *Moss Trooper 2nd* [48] (151).

Produce: 1870, Bessie Lee [181]; 1871, Marjorie [205]; 1872, Rowena [244]; 1873, Lord Cecil [256].

[146] BESSIE BELL.

Calved November 25th, 1867.

Bred by and the property of Mr. Thos. McCrae, Guelph, Co. Wellington, Ont.

Sire Dred [15].
Dam Annie Laurie [82] by Black Jock [18].
2nd d. Empress [50] by *Jock* [10].
3rd d. *Beauty* [11] Imported.

Produce: 1869, Bessie Bell 2nd [179].

COWS.

[147] MARY HAY.

Calved October 30th, 1867.

Bred by and the property of Mr. Thomas McCrae, Guelph, Co. Wellington, Ont.

Sire Dred [15].
Dam *Polly Shaw* [174] Imported.

Produce: 1870, Mary Gray [183]; 1872, Jeannie Deans [243]; 1873, Clandeboye [272]; 1874, Lord Wellington [332]; 1876, Dalrymple [331].

[148] MAGGIE LAUDER.

Calved November 5th, 1867.

Bred by Mr. Thomas McCrae, Guelph, Co. Wellington, Ont.; the property of Mr. J. Giles, Boston, Mass., U. S.

Sire Dred [15].
Dam *Chloe* [4] Imported.

[149] BELLA.

Calved October 20th, 1867.

Bred by and the property of Mr. Thomas McCrae, Guelph, Co. Wellington, Ont.

Sire Dred [15].
Dam Jessie Miller [98] by *Jock* [10].
2nd d. *Maggie* [48] by *Moss Trooper 2nd* [48] (151).

[150] *VICTORIA.*

Calved about 1860.

Imported by Mr. George Roddick, Cobourg, Ont.
Sire *Black Prince* [152].
Dam *Beauty* [151] by *Peter* (116).

Produce: 1863, Jennie [152]; 1865, Young Blucher [164]; 1866, Lord Napier [150].

[151] *BEAUTY.*

Bred by the Earl of Selkirk, Scotland.
Sire *Peter* (116).

Produce: 1875, Galloway Tam [70]; ——, The Earl [69].

[152] JENNIE.

Calved September 22nd, 1863.

The property of Mr. John Coleman, West Flamboro', Co. Wentworth, Ont.

Sire Garibaldi [130].
Dam *Victoria* [150] by *Black Prince* [152].
2nd d. *Beauty* [151] by *Peter* (116).

Produce: 1867, Prince Oscar [151]; 1868, Dagmar [153]; 1870, Lily Dale [233].

[153] DAGMAR.

Calved April 16th, 1868.

Bred by and the property of Mr. John Coleman, West Flamboro', Co. Wentworth, Ont.

Sire Black Jock [18].
Dam Jennie [152] by Garibaldi [130].
2nd d. *Victoria* [150] by *Black Prince* [152].
3rd d. *Beauty* [151] by *Peter* (116).

Produce: 1875, Wellington [404]; 1876, Magnolia [397]; 1878, Prince [394]; 1879, Cetewayo [392]; 1880, Black Dane [641].

[156] MINNEHAHA.

Calved August 24th, 1866.

Bred by Mr. Thomas McCrae, Guelph, Co. Wellington, Ont.; the property of Mr. James Laidlaw, Guelph, Co. Wellington, Ont.

Sire Saginaw [94].
Dam *Polly Shaw* [174] Imported.

[157] SUSAN.

Calved February 20th, 1868.

Bred by and the property of Mr. Arthur McNeil, Vaughan, Co. York, Ont.

Sire *Dryfe* [222].
Dam Bess [125] by Marquis [19].
2nd d. *Chloe* [4] Imported.

Produce: 1870, Jim [199]; 1871, Sis [203]; 1872, Bob [252].

COWS.

[158]   **DAIRYMAID.**

Calved April 12th, 1865.

Bred by and the property of Mr. Arthur McNeil, Vaughan, Co. York, Ont.

Sire *Dryfe* [222].
Dam Bonny [27] by *Jock* [10].
2nd d. *White Bag* [5] Imported.

[160]   **NELLIE.**

Calved April 1st, 1867.

Bred by Mr. Thos. McCrae, Guelph, Co. Wellington, Ont.; the property of Mr. D. McPherson, Puslinch, Co. Wellington, Ont.

Sire Dred [15].
Dam Molly [17] by Black Jock [18].
2nd d. *Chloe* [4] Imported.

[161]   **COODERINA.**

Calved October 20th, 1868.

Bred by and the property of Mr. Thos. McCrae, Guelph, Co. Wellington, Ont.

Sire Dred [15].
Dam Jessie [49] by Black Jock [18].
2nd d. *Maggie* [48] by *Moss Trooper 2nd* [48] (151).

[164]   **LILY DALE.**

Calved November 30th, 1868.

Bred by Mr. Thomas McCrae, Guelph, Co. Wellington, Ont.; the property of Mr. John McCulloch, Port Elgin, Co. Bruce, Ont.

Sire *Pride of the Speed* [159].
Dam Annie Laurie [82] by Black Jock [18].
2nd d. Empress [50] by *Jock* [10].
3rd d. *Beauty* [11] Imported.

90 GALLOWAY HERD BOOK.

[166] POCAHONTAS 2ND.

Calved March 22nd, 1871.

Bred by and the property of Mr. Thos. McCrae, Guelph, Co. Wellington, Ont.

Sire *Pride of the Speed* [159].
Dam Lady Kenmure [140] by Black Jock [18].
2nd d. Pocahontas [60] by Marquis [19].
3rd d. Bonny [27] by *Jock* [10].
4th d. *White Bay* [5] Imported.

[167] DAISY DEANE.

Calved April 14th, 1867.

Bred by and the property of Mr. Wm. Hood, Guelph, Co. Wellington, Ont.

Sire Our John [106].
Dam Blue Bell [235] by William Wallace [67].
2nd d. Queen of Beauty [95] by Black Jock [18].
3rd d. *Black Bess* [6] Imported.

[168] QUEEN OF THE WEST.

Calved January 2nd, 1869.

Bred by and the property of Mr. Wm. Hood, Guelph, Co. Wellington, Ont.

Sire Our John [106].
Dam Empress [50] by *Jock* [10].
2nd d. *Beauty* [11] Imported.

[170] SALL.

Calved in 1862.

Bred by Mr. Alex. Kerr, Westminster, Co. Middlesex, Ont.; the property of Mr. Joseph Charlton, Duncrief, Co. Middlesex, Ont.

Sire Robert Bruce [66].
Dam *Bonny* [1] Imported.

Produce: 1867, Mary [171]; 1868, Black Rachel [172]; 1869, Black Nellie [173]; 1870, Black Wellington [169].

[171] MARY.

Calved in 1867.

Bred by Mr. Alex. Kerr, Westminster, Co. Middlesex, Ont.; the property of Mr. Joseph Charlton, Duncrief, Co. Middlesex, Ont.

Sire Young Blucher [41].
Dam Sall [170] by Robert Bruce [66].
2nd d. *Bonny* [1] Imported.

[172] BLACK RACHEL.

Calved February 19th, 1868.

Bred by and the property of Mr. Joseph Charlton, Duncrief, Co. Middlesex, Ont.

Sire Young Blucher [41].
Dam Sall [170] by Robert Bruce [66].
2nd d. *Bonny* [1] Imported.

Produce: 1871, Young Moss Trooper [245]; 1872, Young Mary [246].

[173] BLACK NELLIE.

Calved January 22nd, 1869.

Bred by and the property of Mr. Joseph Charlton, Duncrief, Co. Middlesex, Ont.

Sire Young Blucher [41].
Dam Sall [170] by Robert Bruce [66].
2nd d. *Bonny* [1] Imported.

[174] *POLLY SHAW*.

Calved in 1860.

Imported by Mr. Thomas McCrae, Guelph, Co. Wellington, Ont., from the herd of Mr. John Graham, Shaw, Lockerbie, Scotland.

Produce: 1864, Heather Bell 4th [137]; 1865, Maggie [120]; 1866, Minnehaha [156]; 1867, Mary Hay [147]; 1869, Wellington [162]; 1870, Argyle [176].

[176] KATE.

Calved June, 1865.

Bred by Mr. Graham, Vaughan, Co. York, Ont.; the property of Mr. Thos. McCrae, Guelph, Co. Wellington, Ont.

Sire Dred [15].
Dam Molly [17] by Black Jock [18].
2nd d. *Chloe* [4] Imported.

[177] EBONY.

Calved in 1863.

Bred by Mr. George Miller, Markham, Co. Ontario, Ont.; the property of Mr. Thomas McCrae, Guelph, Co. Wellington, Ont.

Sire *Jock* [10].

Dam *Blacky* [8] Imported.

Produce: 1866, Bertie [163]; 1868, Eugenie [184].

[178] MINNIE.

Calved May 20th, 1866.

Bred by and the property of Mr. Thos. McCrae, Guelph, Co. Wellington, Ont.

Sire Dred [15].

Dam *Heather Bell* [12] Imported.

Produce: 1869, Lady Heron [180]; 1871, Waverly [204]; 1872, Ivanhoe [242]; 1875, Sir Garnet [267]; 1876, Border Lass [384]; 1877, McGregor [339]; 1878, Heather Bloom [399].

[179] BESSIE BELL 2ND.

Calved July 25th, 1869.

Bred by and the property of Mr. Thos. McCrae, Guelph, Co. Wellington, Ont.

Sire Bonnie Dundee [136].

Dam Bessie Bell [140] by Dred [15].

2nd d. Annie Laurie [82] by Black Jock [18].

3rd d. Empress [50] by *Jock* [10].

4th d. *Beauty* [11] Imported.

[180] LADY HERON.

Calved December 1st, 1869.

Bred by and the property of Mr. Thos. McCrae, Guelph, Co. Wellington, Ont.

Sire Sir John A. [137].

Dam Minnie [178] by Dred [15].

2nd d. *Heather Bell* [12] Imported.

Produce: 1875, Lady Bell [278]; 1876, Lord Douglas [406]; 1877, Disraeli [519]; 1878, Young Solway [411]; 1879, Lady Maxwell [414]; 1880, Bismarck [393]; 1881, Dainty Davy [636].

COWS.

[181]   BESSIE LEE.

Calved January 18th, 1870.

Bred by and the property of Mr. Thos. McCrae, Guelph, Co. Wellington, Ont.

Sire Bonnie Dundee [136].
Dam Grace Darling [145] by Dred [15].
2nd d. Jessie [49] by Black Jock [18].
3rd d. *Maggie* [48] by *Moss Trooper 2nd* [48] (151).

[182]   LADY GORDON.

Calved April 20th, 1870.

Bred by and the property of Mr. Thos. McCrae, Guelph, Co. Wellington, Ont.

Sire Bonnie Dundee [136].
Dam Lady Kenmure [140] by *Jock* [10].
2nd d. Pocahontas [60] by Marquis [19].
3rd d. Bonny [27] by *Jock* [10].
4th d. *White Bag* [5] Imported.

[183]   MARY GRAY.

Calved August 1st, 1870.

Bred by and the property of Mr. Thos. McCrae, Guelph, Co. Wellington, Ont.

Sire Bonnie Dundee [136].
Dam Mary Hay [147] by Dred [15].
2nd d. *Polly Shaw* [174] Imported.

Produce: 1872, Lady Dufferin [275]; 1874, Major Gray [273]; 1876, Lord Starr [334]; 1877, Jock Gray [338].

[184]   EUGENIE.

Calved March 26th, 1868.

Bred by and the property of Mr. Thos. McCrae, Guelph, Co. Wellington, Ont.

Sire Dred [15].
Dam Ebony [177] by *Jock* [10].
2nd d. *Blacky* [8] Imported.

[185] HYENA.

Calved October 23rd, 1867.

Bred by Mr. W. H. Peterson, Woolwich, Co. Waterloo, Ont.; the property of Mr. Wm. Hood, Guelph, Co. Wellington, Ont.

Sire Honest Tom [98].
Dam Mary [35] by Marquis [19].
2nd d. Queen of Beauty [95] by Black Jock [18].
3rd d. *Black Bess* [6] Imported.

[186] LADY MARION.

Calved April, 1862.

Bred by and the property of Mr. James Graham, Vaughan, Co. York, Ont.

Sire Black Jock [18].
Dam *Beauty* [11] Imported.

[187] MINA.

Calved January 9th, 1869.

Bred by Mr. Thos. McCrae, Guelph, Co. Wellington, Ont.; the property of Mr. Wm. Hood, Guelph, Co. Wellington, Ont.

Sire *Pride of the Speed* [159].
Dam Lavina [55] by Black Jock [18].
2nd d. Empress [50] by *Jock* [10].
3rd d. *Beauty* [11] Imported.

[188] ROSY HOOD.

Calved November 7th, 1869.

Bred by and the property of Mr. Wm. Hood, Guelph, Co. Wellington, Ont.

Sire Our John [106].
Dam Maggie Lauder [110] by Black Jock [18].
2nd d. Empress [50] by *Jock* [10].
3rd d. *Beauty* [11] Imported.

Produce: 1871, Robin Hood [349]; 1872, Rosa [350]; 1876, Johnny Hood [354]; 1878, Rosy Hood 2nd [359]; 1879, Rosy Hood 3rd [367]; 1880, Scott [412].

[189] TOPSY WOPSY.

Calved January 11th, 1870.

Bred by and the property of Mr. Wm. Hood, Guelph, Co. Wellington, Ont.

Sire Our John [106].
Dam Idaho [97] by *Jock* [10].
2nd d. Jessie [49] by Black Jock [18].
3rd d. *Maggie* [48] by *Moss Trooper 2nd* [48] (151).

[191] ADAH.

Calved January 10th, 1871.

Bred by and the property of Mr. Wm. Hood, Guelph, Co. Wellington, Ont.

Sire Our John [106].
Dam Mary [111] by Our John [106].
2nd d. Empress [50] by *Jock* [10].
3rd d. *Beauty* [11] Imported.

[192] SALL.

Calved August 1st, 1866.

Bred by Mr. W. H. Peterson, Woolwich, Co. Waterloo, Ont.; the property of Mr. Wm. Hood, Guelph, Co. Wellington, Ont.

Sire Prince Albert [190].
Dam *Black Bess* [6] Imported.

Produce: 1870, Pearl [195]; 1871, Honest Tom [238]; 1873, Wellington Boy [264].

[193] BLACK BESS.

Calved May 20th, 1869.

Bred by Mr. Isaac Burkholder, Waterloo, Co. Waterloo, Ont.; the property of Mr. Wm. Hood, Guelph, Co. Wellington, Ont.

Sire Black Tom [37].
Dam Wandering Nellie [277] by William Wallace [28].
2nd d. Miss Torrence [204] by William Wallace [28].
3rd d. Black Jane [210] by Jacob Benner's Bull [196].
4th d. *Black Bess* [6] Imported.

[194] GIPSY QUEEN 2ND.

Calved March 28th, 1870.

Bred by and the property of Mr. Wm. Hood, Guelph, Co. Wellington, Ont.

Sire Our John [106].
Dam Dido [88] by Victor (122).
2nd d. Magnolia [84] by Black Jock [18].
3rd d. Galloway Lass [42] by *Jock* [10].
4th d. *Jet* [2] Imported.

Produce: 1874, Empress [270]; 1876, Nelly Grey 2nd [396]; 1877, Jenny Lind [448]; 1878, Lady Princess [644]; 1879, Mollie [461]; 1880, Lady Jane [446]; 1881, King William [490]; 1882, Curley [436].

[195] PEARL.

Calved August 7th, 1870.

Bred by and the property of Mr. Wm. Hood, Guelph, Co. Wellington, Ont.

Sire Honest Tom [98].
Dam Sall [192] by Prince Albert [190].
2nd d. *Black Bess* [6] Imported.

[196] BELLE MAHONE.

Calved November 22nd, 1871.

Bred by and the property of Mr. Wm. Hood, Guelph, Co. Wellington, Ont.

Sire Our John [106].
Dam Nelly Grey [109] by Black Jock [18].
2nd d. *Maggie* [48] by *Moss Trooper 2nd* [48] (151).

Produce: 1874, Blue Bell [438].

[197] QUEEN OF BEAUTY 2ND.

Calved December 24th, 1870.

Bred by and the property of Mr. Wm. Hood, Guelph, Co. Wellington, Ont.

Sire Our John [106].
Dam Lady Isabella [100] by Donald [123].
2nd d. *Chloe* [4] Imported.

COWS.

[200] **TOPSY.**

Calved April 16th, 1870.

Bred by and the property of Mr. Thos. McCrae, Guelph, Co. Wellington, Ont.

Sire *Pride of the Speed* [159].
Dam *Chloe* [4] Imported.

[201] **JEAN.**

Calved February 5th, 1870.

Bred by and the property of Mr. Arthur McNeil, Vaughan, Co. York, Ont.

Sire Hardfortune [154].
Dam Agnes [124] by Norval [72].
2nd d. Bess [125] by Marquis [19].
3rd d. *Chloe* [4] Imported.

[202] **FANCY.**

Calved February 25th, 1871.

Bred by and the property of Mr. A. McNeil, Vaughan, Co. York, Ont.
Sire Hardfortune [154].
Dam Woolwich Queen [96] by William Wallace [28].
2nd d. Black Bess [38] by Black Tom [37].
3rd d. *Bell* [9] Imported.

Produce: 1875, Mary [281]; 1876, Badger Boy [296]; 1877, Scottish Chief [300]; 1878, Cherry [307]; 1879, Badger [325]; 1881, Woolwich Queen [378].

[203] **SIS.**

Calved March 3rd, 1871.

Bred by Mr. Arthur McNeil, Vaughan, Co. York, Ont.; the property of Mr. Peter Davy, Ashippun, Co. Dodge, Wis., U.S.
Sire Hardfortune [154].
Dam Susan [157] by *Dryfe* [222].
2nd d. Bess [125] by Marquis [19].
3rd *Chloe* [4] Imported.

Produce: 1874, Nelly Grey [269]; 1875, Isabel [282]; 1876, Dora Bell [297]; 1877, Betty [299]; 1878, Rosa [306]; 1879, Heenan [321]; 1880, Blooming Heather [336]; 1881, Rob [380].

[204]   MISS TORRENCE.

Calved about 1864.

Bred by and the property of Mr. John Torrence, Vaughan, Co. York, Ont.

Sire William Wallace [28].
Dam Black Jane [210] by Jacob Benner's Bull [196].
2nd d. *Black Bess* [6] Imported.

Produce: 1867, Wandering Nellie [277].

[205]   MARJORIE.

Calved February 24th, 1871.

Bred by and the property of Mr. Thos. McCrae, Guelph, Co. Wellington, Ont.

Sire *Pride of the Speed* [159].
Dam Grace Darling [145] by Dred [15].
2nd d. Jessie [49] by Black Jock [18].
3rd d. *Maggie* [48] by *Moss Trooper 2nd* [48] (151).

[206]   QUEEN MAB.

Calved March 1st, 1871.

Bred by and the property of Mr. Thos. McCrae, Guelph, Co. Wellington, Ont.

Sire *Pride of the Speed* [159].
Dam May Queen [143] by Honest Tom [98].
2nd d. Queen of Beauty [95] by Black Jock [18].
3rd d. *Black Bess* [6] Imported.

Produce: 1876, Lord Lochinvar [405]; 1878, Queen Bess [407]; 1879, Black Queen [416]; 1880, Queen Mary [486].

[208]   LIBBIE.

Calved March 3rd, 1867.

Bred by Dr. Bovell, Toronto, Co. York, Ont.; the property of Mr. A. Harrison, Maugerville, New Brunswick.

Sire Dred [15].
Dam Woolwich Queen [96] by William Wallace [28].
2nd d. Black Bess [38] by Black Tom [37].
3rd d. *Bell* [9] Imported.

COWS.

[209]   CRICKET.

Calved March 1st, 1865.

Bred by the late John Snell, Edmonton, Co. Peel, Ont.; the property of Mr. A. Harrison, Maugerville, New Brunswick.
Sire Black Jock [18].
Dam Lavina [55] by Black Jock [18].
2nd d. Empress [50] by *Jock* [10].
2nd d. *Beauty* [11] Imported.

[210]   BLACK JANE.

Calved about 1862.

Breeder unknown.
Sire Jacob Benner's Bull [196].
Dam *Black Bess* [6] Imported.
Produce: 1867, Black Tom [279]; 1864, Miss Torrence [204].

[212]   *BLACK MARIA.*

Calved in 1862.

Owned by Col. R. L. Denison, Dover Court, Toronto, Co. York, Ont.; imported by Mr. Geo. Miller, Markham, Co. York, Ont., from Scotland.

[213]   *TOPSY.*

Calved about 1862.

Owned by Col. R. L. Denison, Dover Court, Toronto, Co. York, Ont.; imported by Mr. Geo. Miller, Markham, Co. York, Ont., from Scotland.
Produce: 1864, Capitola [215]; 1866, Dover Court [216]; 1867, Lippincote [220].

[215]   CAPITOLA.

Calved November 6th, 1864.

Bred by and the property of Col. R. L. Denison, Dover Court, Toronto, Co. York, Ont.
Sire Prince of Wales [57].
Dam *Topsy* [213] Imported.
Produce: 1867, Jane Peterson [221].

[217]   QUEEN'S OWN OF RIDGEWAY.

Calved June 2nd, 1866.

Bred by and the property of Col. R. L. Denison, Dover Court, Toronto, Co. York, Ont.

Sire Wonderful Lad [214].
Dam Selina [104] by Prince of Wales [57].
2nd d. *Helen* [103] Imported.

[218]   BLACK JESSIE.

Calved about 1858.

Bred by and the property of Mr. Graham, Vaughan, Co. York, Ont.;

Sire *Jock* [10].
Dam *Sall* [14] Imported.

Produce: Black Swan [219].

[219]   BLACK SWAN.

Calved about 1861.

Bred by and the property of Mr. John Borer, West Flamboro', Co. Wentworth, Ont.

Sire Prince Albert [190].
Dam Black Jessie [218] by *Jock* [10].
2nd *Sall* [14] Imported.

Produce: White Bag [220].

[220]   WHITE BAG.

Calved about 1865.

Bred by and the property of Mr. John Borer, West Flamboro', Co. Wentworth, Ont.

Sire Moss Trooper 4th [493].
Dam Black Swan [219] by Prince Albert [190].
2nd d. Black Jessie [218] by *Jock* [10].
3rd d. *Sall* [14] Imported.

Produce: Black Swan [219].

COWS.

[221]   JANE PETERSON.

Calved August 1st, 1867.

Bred by and the property of Col. R. L. Denison, Dover Court, Toronto, Co. York, Ont.

         Sire Wonderful Lad [214].
     Dam Capitola [215] by Prince of Wales [57].
       2nd d. *Topsy* [213] Imported.

[223]   BLACK SALL.

Calved January 10th, 1869.

Bred by and the property of Mr. John Borer, West Flamboro', Co. Wentworth, Ont.

         Sire Black Tom [279].
     Dam White Bag [220] by Moss Trooper 4th [493].
     2nd d. Black Swan [219] by Prince Albert [190].
     3rd d. Black Jessie [218] by *Jock* [10].
       4th d. *Sall* [14] Imported.

Produce: 1874, Meg Hood [479]; 1875, Fred [475].

[224]   BEAUTY.

Calved about 1866.

Bred by and the property of Mr. John Borer, West Flamboro', Co. Wentworth, Ont.

         Sire Prince Albert [190].
     Dam Black Swan [219] by Prince Albert [190].
     2nd d. Black Jessie [218] by *Jock* [10].
       3rd d. *Sall* [14] Imported.

Produce: 1873, Gipsy [225].

[225]   GIPSY.

Calved about 1873.

Bred by Mr. H. C. Tew, West Flamboro', Co. Wentworth, Ont.; the property of Mr. Wm. Hood, Guelph, Co. Wellington, Ont.

         Sire Moss Trooper 4th [493].
     Dam Beauty [224] by Prince Albert [190].
     2nd d. Black Swan [219] by Prince Albert [190].
     3rd d. Black Jessie [218] by *Jock* [10].
       4th d. *Sall* [14] Imported.

Produce: 1872, Aggie [259]; 1874, Bismarck [638]; 1876, Kate [296]; 1878, Curly Jock [639].

[233] LILY DALE.

Calved January 29th, 1870.

Bred by Mr. John Coleman, West Flamboro', Co. Wentworth, Ont.; the property of Mr. Wm. Hood, Guelph, Co. Wellington, Ont.

Sire Prince Oscar [151].
Dam Jennie [152] by Garibaldi [130].
2nd d. Victoria [150] by *Black Prince* [152].
3rd d. *Beauty* [151] by *Peter* (116).

Produce: 1873, Robin Hood [256]; 1875, Nelson [435]; 1881, Fairy Dale [482].

[234] GALLOWAY LASS.

Calved May 10th, 1870.

Bred by Mr. John Laidlaw, Guelph, Co. Wellington, Ont.; the property of Mr. Wm. Hood, Guelph, Co. Wellington, Ont.

Sire *Pride of the Speed* [159].
Dam Maümee [92] by *Jock* [10].
2nd *Newbie Lass* [75] Imported.

Produce: 1873, Black Bess [260]; 1874, Johnny Cope [265]; 1876, Wellington Boy [432]; 1877, Rose [385]; 1878, Pocahontas 3rd [487]; 1879, Lorne [447]; 1881, Helen [655].

[235] BLUE BELL.

Calved September 2nd, 1864.

Bred by Mr. W. H. Peterson, Woolwich, Co. Waterloo, Ont.; the property of Mr. Wm. Hood, Guelph, Co. Wellington, Ont.

Sire William Wallace [67].
Dam Queen of Beauty [95] by Black Jock [18].
2nd d. *Black Bess* [6] Imported.

[236] QUEEN VIC.

Calved May 24th, 1872.

Bred by and the property of Mr. Wm. Hood, Guelph, Co. Wellington, Ont.

Sire Robin [166].
Dam Lady Isabella [100] by Donald [123].
2nd d. *Chloe* [4] Imported.

COWS.

[237] MAY QUEEN 2ND.

Calved October 30th, 1871.

Bred by and the property of Mr. Wm. Hood, Guelph, Co. Wellington, Ont.

Sire Robin [166].
Dam Empress [50] by *Jock* [10].
2nd d. *Beauty* [11] Imported.

[241] LADY LISGAR.

Calved June 28th, 1872.

Bred by Mr. Thos. McCrae, Guelph, Co. Wellington, Ont.; the property of Mr. Daniel Steinmetz, Schooneck, Co. Lancaster, Penn., U.S.

Sire King Tom [175].
Dam Lady Kenmure [140] by *Jock* [10].
2nd d. Pocahontas [60] by Marquis [19].
3rd d. Bonny [27] by *Jock* [10].
4th d. *White Bag* [5] Imported.

[243] JEANNIE DEANS.

Calved March 10th, 1872.

Bred by and the property of Mr. Thos. McCrae, Guelph, Co. Wellington, Ont.

Sire King Tom [175].
Dam Mary Hay [147] by Dred [15].
2nd d. *Polly Shaw* [174] Imported.

Produce: 1876, Bonny Bessie [492]; 1878, Laird of Burleigh [516].

[244] ROWENA.

Calved April 5th, 1872.

Bred by and the property of Mr. Thos. McCrae, Guelph, Wellington, Ont.

Sire King Tom [175].
Dam Grace Darling [145] by Dred [15].
2nd d. Jessie [49] by Black Jock [18].
3rd d. *Maggie* [48] by *Moss Trooper 2nd* [48] (151).

[246] YOUNG MARY.

Calved October 12th, 1872.

Bred by and the property of **Mr. Joseph Charlton, Dunciief, Co. Middlesex, Ont.**

Sire Robert Bruce [66].
Dam Black Rachel [172] by Young Blucher [41].
2nd d. Sall [170] by Robert Bruce [66].
3rd d. *Bonny* [1] Imported.

[251] ANNIE LAURIE 2ND.

Calved December 8th, 1872.

Bred by and the property of **Mr. Wm. Dow, Barnett, Co. Wellington, Ont.**

Sire Johnny Cope [165].
Dam May Queen [143] by Honest Tom [98].
2nd d. Queen of Beauty [95] by Black Jock [18].
3rd d. *Black Bess* [6] Imported.

Produce: 1878, Blackamoor [391]; 1879, Mary Little [421]; 1880, Belle of Janefield [338].

[257] JESSIE.

Calved June 22nd, 1871.

Bred by Mr. Isaac Burkholder, Waterloo, Co. Waterloo, Ont.; the property of **Mr. Wm. Hood, Guelph, Co. Wellington, Ont.**

Sire Black Tom [279].
Dam Wandering Nellie [277] by William Wallace [28].
2nd d. Miss Torrence [204] by William Wallace [28].
3rd d. Black Jane [210] by Jacob Benner's Bull [196].
4th d. *Black Bess* [6] Imported.

Produce: 1875, Sam [280]; 1876, Bonny Lass [295]; 1877, Rob Roy [298]; 1878, Captain [308]; 1879, Jessie 2nd [320]; 1880, Maud [337]; 1881, Ben [381].

[258] BEAUTY.

Calved October 10th, 1871.

Bred by Mr. Arthur McNeil, Vaughan, Co. York, Ont.; the property of **Mr. Wm. Hood, Guelph, Co. Wellington, Ont.**

Sire Hardfortune [154].
Dam Young Bonny [39] by Black Tom [37].
2nd d. *Bell* [9] Imported.

Produce: 1873, Annie Sis [262].

[259] AGGIE.

Calved April 10th, 1872.

Bred by and the property of Mr. Henry C. Tew, West Flamboro', Co. Wentworth, Ont.

Sire Black Jock [18].
Dam Gipsy [225] by Moss Trooper 4th [493].
2nd d. Beauty [224] by Prince Albert [190].
3rd d. Black Swan [219] by Prince Albert [190].
4th d. Black Jessie [218] by *Jock* [10].
5th d. *Sall* [14] Imported.

Produce: 1877, Jet [398]; 1880, Nelly Grey 2nd [402].

[260] BLACK BESS.

Calved February 28th, 1873.

Bred by and the property of Mr. Wm. Hood, Guelph, Co. Wellington, Ont.

Sire Robin [166].
Dam Galloway Lass [234] by *Pride of the Speed* [159].
2nd d. Maümee [92] by *Jock* [10].
3rd d. *Newbie Lass* [75] Imported.

[261] PRAIRIE LASS.

Calved March 20th, 1871.

Bred by and the property of Mr. Wm. Hood, Guelph, Co. Wellington, Ont.

Sire Robin [166].
Dam Lucy [502] by Black Jock [18].
2nd d. Jet [343] by Prince Albert [190].
3rd d. *Black Bess* [6] Imported.

Produce: 1876, Mary Hood [383]; 1880, Prairie Flower [415]; 1881, Beauty of Guelph [389].

[262] ANNIE SIS.

Calved August 3rd, 1873.

Bred by and the property of Mr. Wm. Hood, Guelph, Co. Wellington, Ont.

Sire Bob [252].
Dam Beauty [258] by Hardfortune [154].
2nd d. Young Bonny [39] by Black Tom [37].
3rd d. *Bell* [9] Imported.

[268] **WANDERING NELLIE.**

Calved February 8th, 1874.

Bred by and the property of Mr. Wm. Hood, Guelph, Co. Wellington, Ont.

Sire Roger [198].
Dam Lucy [502] by Black Jock [18].
2nd d. Jet [343] by Prince Albert [190].
3rd d. *Black Bess* [6] Imported.

[269] **NELLY GREY.**

Calved March 13th, 1874.

Bred by and the property of Mr. Wm. Hood, Guelph, Co. Wellington, Ont.

Sire Roger [198].
Dam Sis [203] by Hardfortune [154].
2nd d. Susan [157] by *Dryfe* [222].
3rd d. Bess [125] by Marquis [19].
Dam *Chloe* [4] Imported.

[270] **EMPRESS.**

Calved March 16th, 1874.

Bred by and the property of Mr. Wm. Hood, Guelph, Co. Wellington, Ont.

Sire Roger [198].
Dam Gipsy Queen [194] by Our John [106].
2nd d. Dido [88] by Victor [122].
3rd d. Magnolia [84] by Black Jock [18].
4th d. Galloway Lass [42] by *Jock* [10].
5th d. *Jet* [2] Imported.

[274] **LADY HAMILTON.**

Calved April 30th, 1874.

Bred by Mr. Thos. McCrae, Guelph, Co. Wellington, Ont.; the property of Mr. M. R. Platt, Kansas City, Mo., U.S.

Sire Lord Kenmure [271].
Dam Lady Kenmure [140] by *Jock* [10].
2nd d. Pocahontas [60] by Marquis [19].
3rd d. Bonny [27] by *Jock* [10].
Dam *White Bag* [5] Imported.

Produce: 1877, Lady Grey [423]; 1878, Ontario [442]; 1879, Aberdeen [427]; 1880, Passion Flower [422].

COWS.

[275] LADY DUFFERIN.

Calved November 29th, 1872.

Bred by and the property of Mr. Thos. McCrae, Guelph, Co. Wellington, Ont.

Sire King Tom [175].
Dam Mary Gray [183] by Bonnie Dundee [136].
2nd d. Mary Hay [147] by Dred [15].
3rd d. *Polly Shaw* [174] Imported.

Produce : 1876, Lady Gordon [419]; 1877, Blackwood [355]; 1878, Jock a Dink [254]; 1879, Zulu [337]; 1880, Cally [336].

[276] BETTY LAIDLAW.

Bred by Mr. Thos. McCrae, Guelph, Co. Wellington, Ont. ; the property of Mr. Joseph Hickson, Montreal, Que.

Sire Lord Kenmure [271].
Dam Maggie Laidlaw [382] by *Pride of the Speed* [159].
2nd d. Maggie [120] by *Jock* [10].
3rd d. *Polly Shaw* [174] Imported.

Produce : 1876, Matilda [302]; 1877, Black Swan [328]; 1878, Lavinia [329]; 1879, Blanchless [330]; 1880, The Squaw [518].

[277] WANDERING NELLIE.

Calved April 3rd, 1867.

Bred by and the property of Mr. Isaac Burkholder, Waterloo, Co. Waterloo, Ont.

Sire William Wallace [28].
Dam Miss Torrence [204] by William Wallace [28].
2nd d. Black Jane [210] by Jacob Benner's Bull [196].
3rd d. *Black Bess* [6] Imported.

Produce : 1869, Black Bess [193]; 1871, Jessie [267].

[278] LADY BELL.

Calved April 3rd, 1875.

Bred by and the property of Mr. Thos. McCrae, Guelph, Co. Wellington, Ont.

Sire Lord Kenmure [271].
Dam Lady Heron [180] by Sir John A. [45].
2nd d. Minnie [178] by Dred [15].
3rd d. *Heather Bell* [12] Imported.

[281] MARY.

Calved March 14th, 1875.

Bred by and the property of Mr. Peter Davy, Ashippun, Co. Dodge, Wis., U.S.

Sire Bob [252].
Dam Fancy [202] by Hardfortune [154].
2nd d. Woolwich Queen [96] by William Wallace [28].
3rd d. Black Bess [38] by Black Tom [37].
4th *Bell* [9] Imported.

Produce: 1877, Molly [301]; 1878, Tom Scott [305]; 1879, Lucy [324]; 1882, Jeanette [504].

[282] ISABEL.

Calved April 3rd, 1875.

Bred by and the property of Mr. Peter Davy, Ashippun, Co. Dodge, Wis., U.S.

Sire Roger [198].
Dam Sis [203] by Hardfortune [154].
2nd d. Susan [157] by *Dryfe* [222].
3rd d. Bess [125] by Marquis [19].
Dam *Chloe* [4] Imported.

Produce: 1878, Wild Eye [304]; 1879, Pride [322]; 1880, Emperor [342]; 1881, Miss Maud [376]; 1882, Enchanter [546].

[284] BLACK BESS 2ND.

Calved January 10th, 1871.

Bred by Mr. Wm. Hood, Guelph, Co. Wellington, Ont.; the property of Mr. R. G. Hart, Lapeer, Mich., U.S.

Sire Johnny Cope [283].
Dam Margaret [285] by *Dryfe* [222].
2nd d. Black Bess [38] by Black Tom [37].
3rd d. *Bell* [9] Imported.

Produce: 1873, Victoria [290]; 1874, May Queen [292]; 1875, Modock [640].

[285] MARGARET.

Calved March 17th, 1869.

Bred by Mr. Arthur McNeil, Vaughan, Co. York, Ont.; the property of Mr. R. G. Hart, Lapeer, Mich., U.S.

Sire *Dryfe* [222].
Dam Black Bess [38] by Black Tom [37].
2nd d. *Bell* [9] Imported.

Produce: 1871, Black Bess 2nd [284]; 8172, Topsy [287].

COWS.

[286]   DAIRYMAID.

Calved May 20th, 1868.

Bred by Mr. Arthur McNeil, Vaughan, Co. York, Ont.; the property of Mr. R. G. Hart, Lapeer, Mich., U.S.
Sire *Dryfe* [222].
Dam Black Bess [38] by Black Tom [37].
2nd d. *Bell* [9] Imported.

Produce: 1871, Hannah [512]; 1872, Dora [288]; 1873, Buffalo Bill [289]; 1874, Black Rachel [294]; 1876, John Borland [620].

[287]   TOPSY.

Calved January 4th, 1872.

Bred by and the property of Mr. R. G. Hart, Lapeer, Mich., U.S.
Sire Johnny Cope [283].
Dam Margaret [285] by *Dryfe* [222].
2nd d. Black Bess [38] by Black Tom [37].
3rd d. *Bell* [9] Imported.

Produce: 1874, Molly Darling [291]; 1878, Duchess of Sutherland [310].

[288]   DORA.

Calved May 21st, 1872.

Bred by and the property of Mr. R. G. Hart, Lapeer, Mich., U.S.
Sire Shoo-Fly [184].
Dam Dairymaid [286] by *Dryfe* [222].
2nd d. Black Bess [38] by Black Tom [37].
3rd d. *Bell* [9] Imported.

Produce: 1874, Dora 2nd [293].

[290]   VICTORIA.

Calved April 11th, 1873.

Bred by and the property of Mr. R. G. Hart, Lapeer, Mich., U.S.
Sire Shoo-Fly [184].
Dam Black Bess 2nd [284] by Johnny Cope 2nd [283].
2nd d. Margaret [285] by *Dryfe* [222].
3rd d. Black Bess [38] by Black Tom [37].
4th d. *Bell* [9] Imported.

[291] MOLLIE DARLING.

Calved October 10th, 1874.

Bred by and the property of Mr. R. G. Hart, Lapeer, Mich., U. S.
Sire Shoo-Fly [184].
Dam Topsy [287] by Johnny Cope 2nd [283].
2nd d. Margaret [285] by *Dryfe* [222].
3rd d. Black Bess [38] by Black Tom [37].
4th d. *Bell* [9] Imported.
Produce: 1877, Gipsy Queen [311].

[292] MAY QUEEN.

Calved May 1st, 1874.

Bred by and the property of Mr. R. G. Hart, Lapeer, Mich., U.S.
Sire Shoo-Fly [184].
Dam Black Bess 2nd [284] by Johnny Cope 2nd [283].
2nd d. Margaret [285] by *Dryfe* [222].
3rd d. Black Bess [38] by Black Tom [37].
4th d. *Bell* [9] Imported.

[293] DORA 2ND.

Calved February 7th, 1874.

Bred by Mr. R. G. Hart, Lapeer, Mich., U.S.; the property of Mr. Anderson De Witt, Washington Court-House, Ohio, U.S.
Sire Shoo-Fly [184].
Dam Dora [288] by Shoo Fly [184].
3rd d. Dairymaid [286] by *Dryfe* [222].
2nd d. Black Bess [38] by Black Tom [37].
3rd d. *Bell* [9] Imported.
Produce: 1882, La Fayette [642]; 1883, Nell of Wea [658].

[294] BLACK RACHEL.

Calved February 8th, 1874.

Bred by and the property of Mr. R. G. Hart, Lapeer, Mich., U S.
Sire Shoo-Fly [184].
Dam Dairymaid [286] by *Dryfe* [222].
2nd d. Black Bess [38] by Black Tom [37].
3rd d. *Bell* [9] Imported.
Produce: 1876, Black Beauty [309].

COWS.

[295]   BONNY LASS.

Calved March 9th, 1876.

Bred by and the property of Mr. Peter Davy, Ashippun, Co. Dodge, Wis., U.S.

Sire Wellington Boy [264].
Dam Jessie [257] by Black Tom [279].
2nd d. Wandering Nellie [277] by William Wallace [28].
3rd d. Miss Torrence [204] by William Wallace [28].
4th d. Black Jane [210] by Jacob Benner's Bull [196].
5th d. *Black Bess* [6] Imported.

Produce : 1879, Scotch Maid [323]; 1881, Bill [375].

[296]   KATE.

Calved September 22nd, 1876.

Bred by Mr. Wm. Hood, Guelph, Co. Wellington, Ont.; the property of Messrs. McKay & Burleigh, Mechanicsville, Iowa, U.S.

Sire Roger [198].
Dam Gipsy [225] by Moss Trooper 4th [493].
2nd d. Beauty [224] by Prince Albert [190].
3rd d. Black Swan [219] by Prince Albert 190].
4th d. Black Jessie [218] by *Jock* [10].
5th d. *Sall* [14] Imported.

[297]   DORA BELL.

Calved March 4th, 1876.

Bred by and the property of Mr. Peter Davy, Ashippun, Co. Dodge, Wis., U.S.

Sire Wellington Boy [264].
Dam Sis [203] by Hardfortune [154].
2nd d. Susan [157] by *Dryfe* [222].
3rd d. Bess [125] by Marquis [19].
4th d. *Chloe* [4] Imported.

Produce : 1879, Rosa [326]; 1880, Sir William [340]; 1881, Dewdrop [377].

[299] BETTY.

Calved February 1st, 1877.

Bred by and the property of Mr. Peter Davy, Ashippun, Co. Dodge, Wis., U.S.

Sire Wellington Boy [264].
Dam Sis [203] by Hardfortune [154].
2nd d. Susan [157] by *Dryfe* [222].
3rd d. Bess [125] by Marquis [19].
4th d. *Chloe* [4] Imported.

[301] MOLLIE.

Calved April 1st, 1877.

Bred by and the property of Mr. Peter Davy, Ashippun, Co. Dodge, Wis., U.S.

Sire Wellington Boy [264].
Dam Mary [281] by Bob [278].
2nd d. Fancy [266] by Hardfortune [154].
3rd d. Woolwich Queen [96] by William Wallace [28].
4th d. Black Bess [38] by Black Tom [37].
5th d. *Bell* [9] Imported.

[302] MATILDA.

Calved January 10th, 1876.

Bred by and the property of Mr. Joseph Hickson, Montreal, Quebec.

Sire Lord Wellington [332].
Dam Betty Laidlaw [276] by Lord Kenmure [271].
2nd d. Maggie Laidlaw [382] by *Pride of the Speed* [159].
3rd d. Maggie [120] by *Jock* [10].
4th d. *Polly Shaw* [174] Imported.

[304] WILD EYE.

Calved January 2nd, 1878.

Bred by and the property of Mr. Peter Davy, Ashippun, Co. Dodge, Wis., U.S.

Sire Sam [280].
Dam Isabel [282] by Roger [198].
2nd d. Sis [203] by Hardfortune [154].
3rd d. Susan [157] by *Dryfe* [222].
4th d. Bess [125] by Marquis [19].
5th d. *Chloe* [4] Imported.

Produce : 1881, Florence [374].

[306] ROSE.

Calved January 4th, 1878.

Bred by and the property of Mr. Peter Davy, Ashippun, Co. Dodge, Wis., U.S.

Sire Sam [280].
Dam Sis [203] by Hardfortune [154].
2nd d. Susan [157] by *Dryfe* [222].
3rd d. Bess [125] by Marquis [19].
4th d. *Chloe* [4] Imported.

Produce: 1880, Comet [341]; 1882, Comet 2nd [617].

[307] CHERRY.

Calved May 8th, 1878.

Bred by and the property of Mr. Peter Davy, Ashippun, Co. Dodge, Wis., U.S.

Sire Sam [280].
Dam Fancy [202] by Hardfortune [154].
2nd d. Woolwich Queen [96] by William Wallace [28].
3rd d. Black Bess [38] by Black Tom [37].
4th d. *Bell* [9] Imported.

Produce: 1880, Sir Walter [379].

[309] BLACK BEAUTY.

Calved April 12th, 1876.

Bred by Mr. R. G. Hart, Lapeer, Mich., U.S.; the property of Mr. J. B. Sutherland, Petite Cote, Co. Essex, Ont.

Sire Prince Albert 2nd [239].
Dam Black Rachel [294] by Shoo-Fly [184].
2nd d. Dairymaid [287] by *Dryfe* [222].
3rd d. Black Bess [38] by Black Tom [37].
4th d. *Bell* [9] Imported.

Produce: 1878, Snowball [312].

[310] DUCHESS OF SUTHERLAND.

Calved January 27th, 1878.

Bred by Mr. R. G. Hart, Lapeer, Mich., U.S.; the property of Mr. J. B. Sutherland, Petite Cote, Co. Essex, Ont.

Sire Buffalo Bill [289].
Dam Topsy [287] by Johnny Cope [283].
2nd d. Margaret [285] by *Dryfe* [222].
3rd d. Black Bess [38] by Black Tom [37].
4th d. *Bell* [9] Imported.

[311]   GIPSY QUEEN.

Calved December 23rd, 1877.

Bred by Mr. R. G. Hart, Lapeer, Mich., U.S.; the property of Mr. J. B. Sutherland, Petite Cote, Co. Essex, Ont.

Sire Buffalo Bill [289].
Dam Mollie Darling [291] by Shoo-Fly [184].
2nd d. Topsy [287] by Johnny Cope [283].
3rd d. Margaret [285] by *Dryfe* [222].
4th d. Black Bess [38] by Black Tom [37].
5th d. *Bell* [9] Imported.

[314]   SNOWBALL.

Calved February 11th, 1870.

Bred by and the property of the Michigan State Agricultural College, Lansing, Mich., U.S.

Sire Victor [122].
Dam Blooming Heather 2nd [80] by Dred [15].
2nd d. *Blooming Heather* [53] by *Moss Trooper 2nd* [48] (151).
3rd d. *Mary* by *Fergy* (19).
4th d. *Jane* by *Bob*.

Produce: 1873, Snowflake [315]; 1874, Snowdrift [444]; 1877, Snowbloom [316]; 1878, Snovir [318].

[315]   SNOWFLAKE.

Bred by and the property of the Michigan State Agricultural College, Lansing, Mich., U.S.

Sire Johnny Scott [313].
Dam Snowball [314] by Victor [122].
2nd d. Blooming Heather 2nd [80] by Dred [15].
3rd d. *Blooming Heather* [53] by *Moss Trooper 2nd* [48] (151).
4th d. *Mary* by *Fergy* (19).
5th d. *Jane* by *Bob*.

Produce: 1876, Lassie [445]; 1877, Snip [317]; 1878, Snow [319]; 1880, Catharine [339].

COWS.

[316] SNOWBLOOM.

Calved April 20th, 1877.

Bred by and the property of the Michigan State Agricultural College, Lansing, Mich., U.S.

Sire Johnny Scott [313].
Dam Snowball [314] by Victor [122].
2nd d. Blooming Heather 2nd [80] by Dred [15].
3rd d. *Blooming Heather* [53] by *Moss Trooper 2nd* [48] (151).
4th d. *Mary* by *Fergy* (19).
5th d. *Jane* by *Bob*.

[317] SNIP.

Calved March 28th, 1877.

Bred by and the property of the Michigan State Agricultural College, Lansing, Mich., U.S.

Sire Billy McNeil [327].
Dam Snowflake [315] by Victor [122].
2nd d. Blooming Heather 2nd [80] by Dred [15].
3rd d. *Blooming Heather* [53] by *Moss Trooper 2nd* [48] (151).
4th d. *Mary* by *Fergy* (19).
5th d. *Jane* by *Bob*.

Produce: 1880, Snip 2nd [338]; 1881, Snip 3rd [623]; 1882, Jessamine [622].

[318] SNOVIR.

Calved April 26th, 1878.

Bred by and the property of the Michigan State Agricultural College, Lansing, Mich., U.S.

Sire Johnny Scott [313].
Dam Snowball [314] by Victor [122].
2nd d. Blooming Heather 2nd [80] by Dred [15].
3rd d. *Blooming Heather* [53] by *Moss Trooper 2nd* [48] (151).
4th d. *Mary* by *Fergy* (19).
5th d. *Jane* by *Bob*.

Produce: 1881, Louisa Lorne [417]; 1882, Black Dan [491].

[319] SNOW.

Calved April 30th, 1878.

Bred by and the property of the **Michigan State Agricultural College**, Lansing, U.S.

Sire Johnny Scott [313].
Dam Snowflake [315] by Johnny Scott [313].
2nd d. Snowball [314] by Victor [122].
3rd d. Blooming Heather 2nd [80] by Dred [15].
4th d. *Blooming Heather* [53] by *Moss Trooper 2nd* [48] (151).
5th d. *Mary* by *Fergy* (19).
6th d. *Jane* by *Bob*.

Produce: 1881, Johnny Scott 3rd [621]; 1882, Allan [623].

[320] JESSIE 2ND.

Calved January 3rd, 1879.

Bred by and the property of Mr. **Peter Davy**, Ashippun, Co. Dodge, Wis., U.S.

Sire Sam [280].
Dam Jessie [257] by Black Tom [279].
2nd d. Wandering Nellie [277] by William Wallace [28].
3rd d. Miss Torrence [204] by William Wallace [28].
4th d. Black Jane [210] by Jacob Benner's Bull [196].
5th d. *Black Bess* [6] Imported.

Produce: 1882, Curly Nell [544].

[322] PRIDE.

Calved February 25th, 1879.

Bred by and the property of Mr. **Peter Davy**, Ashippun, Co. Dodge, Wis., U.S.

Sire Sam [280].
Dam Isabel [282] by Roger [198].
2nd d. Sis [203] by Hardfortune [154].
3rd d. Susan [157] by *Dryfe* [222].
4th d. Bess [125] by Marquis [19].
5th d. *Chloe* [4] Imported.

[323] SCOTCH MAID.

Calved March 12th, 1879.

Bred by and the property of Mr. Peter Davy, Ashippun, Co. Dodge, Wis., U.S.

Sire Sam [280].
Dam Bonny Lass [295] by Wellington Boy [264].
2nd d. Jessie [257] by Black Tom [279].
3rd d. Wandering Nellie [277] by William Wallace [28].
4th d. Miss Torrence [204] by William Wallace [28].
5th d. Black Jane [210] by Jacob Benner's Bull [196].
6th d. *Black Bess* [6] Imported.

[324] LUCY.

Calved March 15th, 1879.

Bred by and the property of Mr. Peter Davy, Ashippun, Co. Dodge, Wis., U.S.

Sire Sam [280].
Dam Mary [281] by Bob [278].
2nd d. Fancy [202] by Hardfortune [154].
3rd d. Woolwich Queen [96] by William Wallace [28].
4th d. Black Bess [38] by Black Tom [37].
5th d. *Bell* [9] Imported.

[326] ROSA.

Calved April 28th, 1879.

Bred by and the property of Mr. Peter Davy, Ashippun, Co. Dodge, Wis., U.S.

Sire Sam [280].
Dam Dora Bell [297] by Wellington Boy [264].
2nd d. Sis [203] by Hardfortune [154].
3rd. d. Susan [157] by *Dryfe* [222].
4th d. Bess [125] by Marquis [19].
5th d. *Chloe* [4] Imported.

Produce: 1882, Maude [545].

[328] BLACK SWAN.

Calved January 14th, 1877.

Bred by and the property of Mr. Joseph Hickson, Montreal, Que.
Sire Lord Wellington [332].
Dam Betty Laidlaw [276] by Lord Kenmure [271].
2nd d. Maggie Laidlaw [382] by *Pride of the Speed* [159].
3rd d. Maggie [120] by *Jock* [10].
4th d. *Polly Shaw* [174] Imported.

Produce: 1880, Empress [517].

[329] LAVINIA.

Calved January 23rd, 1878.

Bred by and the property of Mr. Joseph Hickson, Montreal, Que.
Sire Lord Wellington [332].
Dam Betty Laidlaw [276] by Lord Kenmure [271].
2nd d. Maggie Laidlaw [382] by *Pride of the Speed* [159].
3rd d. Maggie [120] by *Jock* [10].
4th d. *Polly Shaw* [174] Imported.

[330] BLANKLESS.

Calved May 10th, 1879.

Bred by and the property of Mr. Joseph Hickson, Montreal, Que.
Sire Lord Wellington [332].
Dam Betty Laidlaw [276] by Lord Kenmure [271].
2nd d. Maggie Laidlaw [382] by *Pride of the Speed* [159].
3rd d. Maggie [120] by *Jock* [10].
4th d. *Polly Shaw* [174] Imported.

[334] JULIAND.

Calved August 1st, 1879.

Bred by Mr. Joseph Juliand, Bainbridge, N.Y., U.S.; the property of
Mr. W. B. Montgomery, Starkville, Miss., U.S.
Sire Don Bernado [333].
Dam Victoria [290] by Shoo-Fly [184].
2nd d. Black Bess 2nd [284] by Johnny Cope [283].
3rd d. Margaret [285] by *Dryfe* [222].
4th d. Black Bess [38] by Black Tom [37].
5th d. *Bell* [9] Imported.

[335] TOPSY.

Calved January 18th, 1876.

Bred by Mr. R. G. Hart, Lapeer, Mich., U.S.; the property of Mr. W. B. Montgomery, Starkville, Miss., U. S.

Sire Buffalo Bill [289].
Dam Victoria [290] by Shoo-Fly [184].
2nd d. Black Bess 2nd [284] by Johnny Cope [283].
3rd. d. Margaret [285] by *Dryfe* [222].
4th d. Black Bess [38] by Black Tom [37].
5th d. [9] *Bell* Imported.

[336] BLOOMING HEATHER.

Calved February 27th, 1880.

Bred by and the property of Mr. Peter Davy, Ashippun, Co. Dodge, Wis., U.S.

Sire Sam [280].
Dam Sis [203] by Hardfortune [154].
2nd d. Susan [157] by *Dryfe* [222].
3rd d. Bess [125] by Marquis [19].
4th d. *Chloe* [4] Imported.

[337] MAUDE.

Calved April 25th, 1880.

Bred by and the property of Mr. Peter Davy, Ashippun, Co. Dodge, Wis, U.S.

Sire Sam [280].
Dam Jessie [257] by Black Tom [279].
2nd d. Wandering Nellie [277] by William Wallace [28].
3rd d. Miss Torrence [204] by William Wallace [28].
4th d. Black Jane [210] by Jacob Benner's Bull [196].
5th d. *Black Bess* [6] Imported.

[338] SNIP 2ND.

Calved February 9th, 1880.

Bred by the Michigan State Agricultural College, Lansing, Mich., U.S.; the property of Mr. Peter Davy, Ashippun, Co. Dodge, Wis., U.S.

Sire Johnny Scott [313].
Dam Snip [317] by Billy McNeil [327].
2nd d. Snowflake [315] by Johnny Scott [313].
3rd d. Snowball [314] by Victor [122].
4th d. Blooming Heather 2nd [80] by Dred [15].
5th d. *Blooming Heather* [53] by *Moss Trooper 2nd* [48] (151).
6th d. *Mary* by *Fergy* (19).
7th d. *Jane* by *Bob*.

[339] CATHARINE.

Calved February 11th, 1880.

Bred by the Michigan State Agricultural College, Lansing, Mich., U.S; the property of Mr. Peter Davy, Ashippun, Co. Dodge, Wis., U.S.

Sire Johnny Scott [313].
Dam Snowflake [315] by Johnny Scott [313].
2nd d. Snowball [314] by Victor [122].
3rd d. Blooming Heather 2nd [80] by Dred [15].
4th d. *Blooming Heather* [53] by *Moss Trooper 2nd* [48] (151).
5th d. *Mary* by *Fergy* (19).
6th d. *Jane* by *Bob*.

[343] JET.

Calved July 3rd, 1865.

Bred by Mr. H. W. Peterson, Woolwich, Co. Waterloo, Ont.; at one time the property of Mr. Wm. Hood, Guelph, Co. Wellington, Ont.; then of Mr. J. N. Smith, Bath, Mich., U.S.; and now in the possession of Mr. R. B. Caruss, St. John, Co. Clinton, Mich., U.S.

Sire Prince Albert [190].
Dam *Black Bess* [6] Imported.

Produce: 1870, Lucy [502]; 1871, Jettie [348]; 1874, Jet 2nd [352]; 1877, Jet 3rd [358].

COWS.

[344] MARGARET.

Calved April, 1868.

Bred by Mr. Arthur McNeil, Vaughan, Co. York, Ont. ; the property of Mr. A. B. Caruss, St. John, Co. Clinton, Mich., U.S.

Sire Hardfortune [154].

Dam Lizzie [114] by Dred [15].

2nd d. Bonny 2nd [74] by *Jock* [10].

3rd d. *Chloe* [4] Imported.

Produce ; 1877, Bob Wooley [410]; 1881, Tom [395].

[345] ROSA McNEIL.

Calved April, 1869.

Bred by Mr. Arthur McNeil, Vaughan, Co. York, Ont. ; the property of Mr. R. B. Caruss, St. John, Co. Clinton, Mich., U.S.

Sire Hardfortune [154].

Dam Lizzie [114] by Dred [15].

2nd d. Bonny 2nd [74] by *Jock* [10].

3rd d. *Chloe* [4] Imported.

Produce : 1873, McNeil [443]; 1875, Billy McNeil [327]; 1879, Johnny McNeil [366]; 1880, Topsy [369]; 1881, Rosy [510].

[347] MAGGIE LAUDER 2ND.

Calved March 20th, 1870.

Bred by Mr. Wm. Hood, Guelph, Co. Wellington, Ont. ; the property of Mr. J. N. Smith, Bath, Mich., U.S.

Sire Our John [106].

Dam Maggie Lauder [110] by Black Jock [18].

2nd d. Empress [50] by *Jock* [10].

3rd d. *Beauty* [11] Imported.

Produce : 1878, Maggie 5th [360]; 1881, Miss Lauder [488].

[348] JETTIE.

Calved May 20th, 1871.

Bred by Mr. Wm. Hood, Guelph, Co. Wellington, Ont. ; the property of Mr. M. R. Platt, Kansas City, Mo., U.S.

Sire Our John [106].

Dam Jet [343] by Prince Albert [190].

2nd d. *Black Bess* [6] Imported.

Produce : 1874, Lady Black [648]; 1877, Jettie 2nd [357]; 1878, Jettie 3rd [361]; 1880, Nora [372].

[350] ROSA.

Calved May 10th, 1872.

Bred by and the property of Mr. J. N. Smith, Bath, Mich., U.S.

Sire Napier [346].

Dam Rosy Hood [188] by Our John [106].

2nd d. Maggie Lauder [110] by Black Jock [18].

3rd d. Empress [50] by *Jock* [10].

4th d. *Beauty* [11] Imported.

Produce: 1876, Rosa 2nd [355]; 1879, Rosy Hood 4th [368].

[352] JET 2ND.

Calved April 15th, 1874.

Bred by Mr. J. N. Smith, Bath, Co. Clinton, Mich., U.S.; the property of Mr. R. B. Caruss, St. John, Co. Clinton, Mich., U.S.

Sire Robin Hood [349].

Dam Jet [343] by Prince Albert [190].

2nd d. *Black Bess* [6] Imported.

Produce: 1878, Jet 4th [362]; 1880, Nero 2nd [371].

[353] MAGGIE LAUDER 3RD.

Calved April 10th, 1875.

Bred by Mr. J. N. Smith, Bath, Mich., U.S.; the property of Mr. R. B. Caruss, St. John, Co. Clinton, Mich., U.S.

Sire Robin Hood [349].

Dam Maggie Lauder [110] by Black Jock [18].

2nd d. Empress [50] by *Jock* [10].

3rd d. *Beauty* [11] Imported.

Produce: 1877, Blackbird [633]; 1879, Maggie 3rd [364].

[355] ROSA 2ND.

Calved August 5th, 1876.

Bred by and the property of Mr. J. N. Smith, Bath, Co. Clinton, Mich., U.S.

Sire Robin Hood [349].

Dam Rosa [350] by Napier [346].

2nd d. Rosy Hood [188] by Our John [106].

3rd d. Maggie Lauder [110] by Black Jock [18].

4th d. Empress [50] by *Jock* [10].

5th d. *Beauty* [11] Imported.

Produce: 1879, Rosa 3rd [363].

COWS.

[356]   MAGGIE 4TH.

Calved July 1st, 1877.

Bred by Mr. J. N. Smith, Bath, Co. Clinton, Mich., U.S.; the property of Mr. M. R. Platt, Kansas City, Mo., U.S.
Sire Robin Hood [349].
Dam Maggie Lauder [110] by Black Jock [18].
2nd d. Empress [50] by *Jock* [10].
3rd d. *Beauty* [11] Imported.

[357]   JETTIE 2ND.

Calved June 10th, 1877.

Bred by Mr. J. N. Smith, Bath, Co. Clinton, Mich., U.S.; the property of Mr. M. R. Platt, Kansas City, Mo., U.S.
Sire Robin Hood [349].
Dam Jettie [348] by Our John [106].
2nd d. Jet [343] by Prince Albert [190].
3rd d. *Black Bess* [6] Imported.

[358]   JET 3RD.

Calved June 2nd, 1877.

Bred by Mr. J. N. Smith, Bath, Co. Clinton, Mich., U.S.; the property of Mr. R. B. Caruss, St. John, Co. Clinton, Mich., U.S.
Sire Robin Hood [349].
Dam Jet [343] by Prince Albert [190].
2nd d. *Black Bess* [6] Imported.

Produce: 1879, Sally [365]; 1880, Independence [370]; 1881, Governor St. John [413].

[359]   ROSY HOOD 2ND.

Calved June 6th, 1878.

Bred by Mr. J. N. Smith, Bath, Co. Clinton, Mich., U.S.; the property of Mr. M. R. Platt, Kansas City, Mo., U.S.
Sire Billy McNeil [327].
Dam Rosy Hood [188] by Our John [106].
2nd d. Maggie Lauder [110] by Black Jock [18].
3rd d. Empress [50] by *Jock* [10].
4th d. *Beauty* [11] Imported.

Produce: 1880, Nero [373].

[360]   MAGGIE 5th.

Calved August 4th, 1878.

Bred by Mr. J. N. Smith, Bath, Co. Clinton, Mich., U.S.; the property of Mr. M. R. Platt, Kansas City, Mo., U.S.

Sire Billy McNeil [327].
Dam Maggie Lauder 2nd [347] by Our John [106].
2nd d. Maggie Lauder [110] by Black Jock [18].
3rd d. Empress [50] by *Jock* [10].
4th d. *Beauty* [11] Imported.

361]   JETTIE 3rd.

Calved June 1st, 1878.

Bred by Mr. J. N. Smith, Bath, Co. Clinton, Mich., U.S.; the property of Mr. M. R. Platt, Kansas, City, Mo., U.S.

Sire Billy McNeil [327].
Dam Jettie [348] by Our John [106].
2nd d. Jet [343] by Prince Albert [190].
3rd d. *Black Bess* [6] Imported.

Produce: 1881, John Brown [429].

[362]   JET 4th.

Calved March 6th, 1878.

Bred by Mr. J. N. Smith, Bath, Co. Clinton, Mich., U.S.; the property of Mr. R. B. Caruss, St. John, Co. Clinton, Mich., U.S.

Sire Billy McNeil [327].
Dam Jet 2nd [352] by Robin Hood [349].
2nd d. Jet [343] by Prince Albert [190].
3rd d. *Black Bess* [6] Imported.

[363]   ROSA 3rd.

Calved May 5th, 1879.

Bred by and the property of Mr. J. N. Smith, Bath, Co. Clinton, Mich., U.S.

Sire Billy McNeil [327].
Dam Rosa 2nd [355] by Robin Hood [349].
2nd d. Rosa [350] by Napier [346].
3rd d. Rosy Hood [188] by Our John [106].
4th d. Maggie Lauder [110] by Black Jock [18].
5th d. Empress [50] by *Jock* [10].
6th d. *Beauty* [11] Imported.

COWS.

[364]   MAGGIE 3RD.

Calved April 12th, 1879.

Bred by and the property of Mr. R. B. Caruss, St. John, Co. Clinton, Mich., U.S.

                 Sire Johnny Hood [354].
       Dam Maggie Lauder 3rd [353] by Robin Hood [349].
         2nd d. Maggie Lauder [110] by Black Jock [18].
             3rd d. Empress [50] by *Jock* [10].
               4th d. *Beauty* [11] Imported.

Produce: 1882, Maggie 4th [599].

[365]   SALLY.

Calved March 10th, 1879.

Bred by and the property of Mr. R. B. Caruss, St. John, Co. Clinton, Mich., U.S.

                 Sire Johnny Hood [354].
          Dam Jet 3rd [358] by Robin Hood [349].
           2nd d. Jet [343] by Prince Albert [190].
              3rd d. *Black Bess* [6] Imported.

Produce: 1882, Lochinvar, Jr. [508].

[367]   ROSY HOOD 3RD.

Calved May 20th, 1879.

Bred by Mr. J. N. Smith, Bath, Co. Clinton, Mich., U.S.; the property of Mr. M. R. Platt, Kansas City, Mo., U.S.

                 Sire Billy McNeil [327].
           Dam Rosy Hood [188] by Our John [106].
         2nd d. Maggie Lauder [110] by Black Jock [18].
             3rd d. Empress [50] by *Jock* [10].
               4th d. *Beauty* [11] Imported.

[368]   ROSY HOOD 4TH.

Calved June 2nd, 1879.

Bred by Mr. J. N. Smith, Bath, Co. Clinton, Mich., U.S; the property of Mr. M. R. Platt, Kansas City, Mo., U.S.

                 Sire Billy McNeil [327].
             Dam Rosa [354] by Napier [346].
          2nd d. Rosy Hood [188] by Our John [106].
        3rd d. Maggie Lauder [110] by Black Jock [18].
            4th d. Empress [50] by *Jock* [10].
              5th d. *Beauty* [11] Imported.

[369]   TOPSY.

Calved April 20th, 1880.

Bred by and the property of Mr. R. B. Caruss, St. John, Co. Clinton, Mich., U.S.

Sire Johnny Hood [354].
Dam Rosy McNeil [345] by Hardfortune [154].
2nd d. Lizzie [114] by Dred [15].
3rd d. Bonny 2nd [74] by *Jock* [10].
4th d. *Chloe* [4] Imported.

Produce: 1882, Topsy 2nd [684].

[372]   NORA.

Calved March 5th, 1880.

Bred by Mr. J. N. Smith, Bath, Co. Clinton, Mich., U.S.; the property of Mr. James McDaniels, Bath, Co. Clinton, Mich., U.S.

Sire Billy McNeil [327].
Dam Jettie [348] by Our John [106].
2nd d. Jet [343] by Prince Albert [190].
3rd d. *Black Bess* [6] Imported.

[374]   FLORENCE.

Calved May 15th, 1881.

Bred by and the property of Mr. Peter Davy, Ashippun, Co. Dodge, Wis., U.S.

Sire Johnny Scott [313].
Dam Wild Eye [304] by Sam [280].
2nd d. Isabel [282] by Roger [198].
3rd d. Sis [203] by Hardfortune [154].
4th d. Susan [157] by *Dryfe* [222].
5th d. Bess [125] by Marquis [19].
6th d. *Chloe* [4] Imported.

[376]   MISS MAUDE.

Calved April 14th, 1881.

Bred by and the property of Mr. Peter Davy, Ashippun, Co. Dodge, Wis., U.S.

           Sire Johnny Scott [313]
Dam Isabel [282] by Roger [198].
2nd d. Sis [203] by Hardfortune [154].
3rd d. Susan [157] by *Dryfe* [222].
4th d. Bess [125] by Marquis [19].
2nd d. *Chloe* [4] Imported.

[377]   DEW DROP.

Calved February 1st, 1881.

Bred by and the property of Mr. Peter Davy, Ashippun, Co. Dodge, Wis., U.S.

           Sire Johnny Scott [313].
Dam Dora Bell [297] by Wellington Boy [264].
2nd d. Sis [203] by Hardfortune [154].
3rd d. Susan [157] by *Dryfe* [222].
4th d. Bess [125] by Marquis [19].
5th d. *Chloe* [4] Imported.

[378]   WOOLWICH QUEEN.

Calved April 15th, 1881.

Bred by and the property of Mr. Peter Davy, Ashippun, Co. Dodge, Wis., U.S.

           Sire Johnny Scott [313].
Dam Fancy [202] by Hardfortune [154].
2nd d. Woolwich Queen [96] by William Wallace [28].
3rd d. Black Bess [38] by Black Tom [37].
4th d. *Bell* [9] Imported.

[382]   MAGGIE LAIDLAW.

Calved June 20th, 1869.

Bred by Mr. Thomas McCrae, Guelph, Co. Wellington, Ont.; the property of Mr. A. B. Matthews, Kansas City, Mo., U.S.

           Sire *Pride of the Speed* [159].
Dam Maggie [120] by *Jock* [10].
2nd d. *Polly Shaw* [174] Imported.

Produce: 1872, Pride [520]; 1874, Betty Laidlaw [276]; 1875, The Warden [263]; 1878, Lord Chelmsford [521]; 1879, Maggie Shaw [476].

[383] MARY HOOD.

Calved April 10th, 1876.

Bred by Mr. Wm. Hood, Guelph, Co. Wellington, Ont.; the property of Mr. A. B. Matthews, Kansas City, Mo., U.S.

Sire Black Jock 2nd [253].
Dam Prairie Lass [261] by Robin [166].
2nd d. Lucy [502] by Black Jock [18].
3rd d. Jet [343] by Prince Albert [190].
4th d. *Black Bess* [6] Imported.

Produce: 1878, Border Belle [522]; 1879, Fanny Blair [477].

[384] BORDER LASS.

Calved June 2nd, 1876.

Bred by Mr. Wm. McCrae, Manor Bank, Guelph, Co. Wellington, Ont.; the property of Mr. A. B. Matthews, Kansas City, Mo., U.S.

Sire *Young Lochinvar* [303].
Dam Minnie [178] by Dred [15].
2nd d. *Heather Bell* [12] Imported.

Produce: 1881, Peter Stewart [390].

[385] ROSE.

Calved February 6th, 1877.

Bred by Mr. Wm. Hood, Guelph, Co. Wellington, Ont.; the property of Mr. A. B. Matthews, Kansas City, Mo., U.S.

Sire Black Jock 2nd [253].
Dam Galloway Lass [234] by *Pride of the Speed* [159].
2nd d. Maümee [92] by *Jock* [10].
3rd d. *Newbie Lass* [75] Imported.

Produce: 1879, Robbie Burns [386]; 1880, Lily [409]; 1881, Doctor Hornbook [387].

[388] BELLE OF JANEFIELD.

Calved December 4th, 1880.

Bred by Mr. Thos. McCrae, Guelph, Co. Wellington, Ont.; the property of Mr. A. B. Matthews, Kansas City, Mo., U.S.

Sire Lord Lochinvar [405].
Dam Annie Laurie 2nd [251] by Johnny Cope [165].
2nd d. May Queen [143] by Honest Tom [98].
3rd d. Queen of Beauty [95] by Black Jock [18].
4th d. *Black Bess* [6] Imported.

COWS.

[389]   **BEAUTY OF GUELPH.**
Calved January 1st, 1881.

Bred by Mr. Thomas McCrae, Guelph, Co. Wellington, Ont.; the property of Mr. A. B. Matthews, Kansas City, Mo., U.S.

Sire Lord Lochinvar [405].
Dam Prairie Lass [261] by Robin [166].
2nd d. Lucy [502] by Black Jock [18].
3rd d. Jet [343] by Prince Albert [190].
4th d *Black Bess* [6] Imported.

[396]   **NELLY GREY 2ND.**
Calved April 8th, 1876.

Bred by Mr. Wm. Hood, Guelph, Co. Wellington, Ont.; the property of Mr. A. B. Matthews, Kansas City, Mo., U.S.

Sire Black Jock 2nd [253].
Dam Gipsy Queen [194] by Our John [106].
2nd d. Dido [88] by Victor [122].
3rd d. Magnolia [84] by Black Jock [18].
4th d. Galloway Lass [42] by *Jock* [10].
5th d. *Jet* [2] Imported.

[397]   **MAGNOLIA 2ND.**
Calved May 17th, 1876.

Bred by Mr. Geo. Hood, Guelph, Co. Wellington, Ont.; the property of Mr. A. B. Matthews, Kansas City, Mo., U.S.

Sire Black Jock 2nd [253].
2nd d. Dagmar [153] by Black Jock [18].
3rd d. Jennie [152] by Garibaldi [130].
4th d. Victoria [150] by *Black Prince* [152].
5th d. *Beauty* [151] by *Peter* (116).

[398]   **JET.**
Calved January 24th, 1877.

Bred by Mr. Wm. Hood, Guelph, Co. Wellington, Ont.; the property of Mr. A. B. Matthews, Kansas City, Mo., U.S.

Sire Black Jock 2nd [253].
Dam Aggie [259] by Black Jock [18].
2nd d. Gipsy [225] by Moss Trooper 4th [493].
3rd d. Beauty [224] by Prince Albert [190].
4th d. Black Swan [219] by Prince Albert [190].
5th d. Black Jessie [218] by *Jock* [10].
6th d. *Sall* [14] Imported.

Produce: 1880, **Jet 2nd** [403].

[399] HEATHER BLOOM.

Calved October 28th, 1878.

Bred by Mr. Wm. McCrae, Manor Bank, Guelph, Co. Wellington, Ont.;
the property of Mr. A. B. Matthews, Kansas City, Mo., U.S.
Sire Lord Lochinvar [405].
Dam Minnie [178] by Dred [15].
2nd d. *Heather Bell* [12] Imported.

[400] LITTLE MARY.

Calved November 19th, 1878.

Bred by Mr. Wm. McCrae, Manor Bank, Guelph, Co. Wellington, Ont.;
the property of Mr. A. B. Matthews, Kansas City, Mo., U.S.
Sire Lord Lochinvar [405].
Dam Mary [111] by Our John [106].
2nd d. Empress [50] by *Jock* [10].
3rd. d. *Beauty* [11] Imported.

[401] MAGNOLIA 3RD.

Calved March 10th, 1880.

Bred by Mr. Henry Devlin, Guelph, Co. Wellington, Ont.; the property of Mr. A. B. Matthews, Kansas City, Mo., U.S.
Sire Wellington 2nd [404].
Dam Magnolia [84] by Black Jock [
2nd d. Galloway Lass [42] by *Jock* [10].
3rd d. *Jet* [2] Imported.

[402] NELLY GREY 3

Calved August 10th, 1880.

Bred by Mr. Henry Devlin, Guelph, Co. Wellington, Ont.; the property of Mr. A. B. Matthews, Kansas City, Mo., U.S.
Sire Wellington 2nd [404].
Dam Aggie [259] by Black Jock [18].
2nd d. Gipsy [225] by Moss Trooper 4th [493].
3rd d. Beauty [224] by Prince Albert [190].
4th d. Black Swan [219] by Prince Albert [190].
5th d. Black Jessie [218] by *Jock* [10].
6th d. *Sall* [14] Imported.

COWS.

[403]   JET 2ND.

Calved October 27th, 1880.

Bred by Mr. Henry Devlin, Guelph, Co. Wellington, Ont.; the property of Mr. A. B. Matthews, Kansas City, Mo., U.S.

Sire Wellington 2nd [404].
Dam Jet [398] by Black Jock 2nd [253].
2nd d. Aggie [259] by Black Jock [18].
3rd d. Gipsy [225] by Moss Trooper 4th [493].
4th d. Beauty [224] by Prince Albert [190].
5th d. Black Swan [219] by Prince Albert [190].
6th d. Black Jessie [218] by *Jock* [10].
7th d. *Sall* [14] Imported.

[407]   QUEEN BESS.

Calved January 5th, 1878.

Bred by and the property of Mr. Wm. McCrae, Manor Bank, Guelph, Co. Wellington, Ont.

Sire Major Gray [273].
Dam Queen Mab [206] by *Pride of the Speed* [159].
Dam May Queen [143] by Honest Tom [98].
2nd d. Queen of Beauty [95] by Black Jock [18].
3rd d. *Black Bess* [6] Imported.

[408]   BONNIE LASSIE.

Calved October 14th, 1879.

Bred by Mr. Thos. McCrae, Guelph, Co. Wellington, Ont.; the property of Mr. A. B. Matthews, Kansas City, Mo., U.S.

Sire Lord Lochinvar [405].
Dam Bonnie Bessie [492] by Lord Kenmure [271].
2nd d. Jeannie Deans [243] by King Tom [175].
3rd d. Mary Hay [147] by Dred [15].
4th d. *Polly Shaw* [174] Imported.

[409]   LILY.

Calved October 27th, 1880.

Bred by Mr. Thos. McCrae, Guelph, Co. Wellington, Ont.; the property of Mr. A. B. Matthews, Kansas City, Mo., U.S.

Sire Lord Douglas [406].
Dam Rose [385] by Black Jock 2nd [253].
2nd d. Galloway Lass [234] by *Pride of the Speed* [159].
3rd d. Maümee [92] by *Jock* [10].
4th d. *Newbie Lass* [75] Imported.

132 GALLOWAY HERD BOOK.

[414] LADY MAXWELL.
Calved July 28th, 1879.

Bred by Mr. Thos. McCrae, Guelph, Co. Wellington, Ont.; the property of Mr. M. R. Platt, Kansas City, Mo., U.S.
Sire Lord Lochinvar [405].
Dam Lady Heron [180] by Sir John A. [137].
2nd d. Minnie [178] by Dred [15].
3rd d. *Heather Bell* [12] Imported.

[415] PRAIRIE FLOWER.
Calved March 12th, 1880.

Bred by and the property of Mr. Thos. McCrae, Guelph, Co. Wellington, Ont.
Sire Lord Douglas [406].
Dam Prairie Lass [261] by Robin [166].
2nd d. Lucy [502] by Black Jock [18].
3rd d. Jet [343] by Prince Albert [190].
4th d. *Black Bess* [6] Imported.

[416] BLACK QUEEN.
Calved January 18th, 1879.

Bred by Mr. Thos. McCrae, Guelph, Co. Wellington, Ont.; the property of Mr. M. R. Platt, Kansas City, Mo., U.S.
Sire Major Gray [273].
Dam Queen Mab [206] by *Pride of the Speed* [159].
2nd d. May Queen [143] by Honest Tom [98].
3rd d. Queen of Beauty [95] by Black Jock [18].
4th d. *Black Bess* [6] Imported.

[417] LOUISE LORNE.
Calved March 19th, 1881.

Bred by and the property of Messrs. Palmer & Sons, Boscobel, Wis., U.S.
Sire Sam [280].
Dam Snovir [318] by Johnny Scott [313].
2nd d. Snowball [314] by Victor [122].
3rd d. Blooming Heather 2nd [80] by Dred [15].
4th d. *Blooming Heather* [53] by *Moss Trooper 2nd* [48] (151).
5th d. *Mary* by *Fergy* (19).
6th d. *Jane* by *Bob*.

[418] QUEEN.

Calved April 10th, 1878.

Bred by Mr. Jacob Weiste, Co. Lancaster, Pa., U.S.; the property of Mr. M. R. Platt, Kansas City, Mo., U.S.

Sire Wellington Boy [432].
Dam Sleeping Maggie [433] by Bob [252].
2nd d. May Queen [143] by Honest Tom [98].
3rd d. Queen of Beauty [95] by Black Jock [18].
4th d. *Black Bess* [6] Imported.

Produce: 1881, Silver King [439].

[419] LADY GORDON.

Calved April 15th, 1876.

Bred by Mr. Thos. McCrae, Guelph, Co. Wellington, Ont.; the property of Mr. M. R. Platt, Kansas City, Mo., U.S.

Sire Lord Kenmure [271].
Dam Lady Dufferin [275] by King Tom [175].
2nd d. Mary Gray [183] by Bonnie Dundee [136].
3rd d Mary Hay [147] by Dred [15].
4th d. *Polly Shaw* [174] Imported.

Produce: 1878, King Jacob [428]; 1879, Hopewell [441]; 1880, Uncle Joe [425]; 1881, Beauty [420].

[420] BEAUTY.

Calved March 2nd, 1881.

Bred by Mr. Jacob Weiste, Co. Lancaster, Pa., U.S.; the property of Mr. M. R. Platt, Kansas City, Mo., U.S.

Sire Wellington Boy [432].
Dam Lady Gordon [419] by Lord Kenmure [271].
2nd d. Lady Dufferin [275] by King Tom [175].
3rd d. Mary Gray [183] by Bonnie Dundee [136].
4th d. Mary Hay [147] by Dred [15].
5th d. *Polly Shaw* [174] Imported.

[421]   MARY LITTLE.

Calved November 15th, 1879.

Bred by and the property of Mr. Thos. McCrae, Guelph, Co. Wellington, Ont.

Sire Lord Douglas [406].
Dam Annie Laurie 2nd [251] by Johnny Cope [165].
2nd d. May Queen [143] by Honest Tom [98].
3rd d. Queen of Beauty [95] by Black Jock [18].
4th d. *Black Bess* [6] Imported.

[422]   PASSION FLOWER.

Calved July 6th, 1880.

Bred by Mr. Jacob Weiste, Co. Lancaster, Pa., U.S.; the property of Mr. M. R. Platt, Kansas City, Mo., U.S.

Sire Wellington Boy [432].
Dam Lady Hamilton [274] by Lord Kenmure [271].
2nd d. Lady Kenmure [140] by *Jock* [10].
3rd d. Pocahontas [60] by Marquis [19].
4th d. Bonny [27] by *Jock* [10].
5th d. *White Bag* [5] Imported.

[423]   LADY GRAY.

Calved February 10th, 1877.

Bred by Mr. Thos. McCrae, Guelph, Co. Wellington, Ont.; the property of Mr. M. R. Platt, Kansas City, Mo., U.S.

Sire Major Gray [273].
Dam Lady Hamilton [274] by Lord Kenmure [271].
2nd d. Lady Kenmure [140] by *Jock* [10].
3rd d. Pocahontas [60] by Marquis [19].
4th d. Bonny [27] by *Jock* [10].
5th d. *White Bag* [5] Imported.

Produce: 1880, Bright Eye [424].

[424] BRIGHT EYE.

Calved May 30th, 1880.

Bred by Mr. Jacob Weiste, Co. Lancaster, Pa., U.S.; the property of Mr. M. R. Platt, Kansas City, Mo., U.S.

Sire Wellington Boy [432].
Dam Lady Gray [423] by Major Gray [273].
2nd d. Lady Hamilton [274] by Lord Kenmure [271].
Dam Lady Kenmure [140] by *Jock* [10].
2nd d. Pocahontas [60] by Marquis [19].
3rd d. Bonny [27] by *Jock* [10].
4th d. *White Bag* [5] Imported.

[433] SLEEPING MAGGIE.

Calved March 30th, 1875.

Bred by and the property of Mr. Wm. Dow, Fergus, Co. Wellington, Ont.

Sire Bob [252].
Dam May Queen [143] by Honest Tom [98].
2nd d. Queen of Beauty [95] by Black Jock [18].
3rd d. *Black Bess* [6] Imported.

Produce: 1878, Queen [418]; 1879, Galloway Boy [426]; 1880, Uncle Sam [440].

[434] LUCY.

Calved June 27th, 1881.

Bred by Mr. Jacob Weiste, Co. Lancaster, Pa., U.S.; the property of Mr. M. R. Platt, Kansas City, Mo., U.S.

Sire Hopewell [441].
Dam Ontario [442] by Wellington Boy [432].
2nd d. Lady Hamilton [274] by Lord Kenmure [271].
3rd d. Lady Kenmure [140] by *Jock* [10].
4th d. Pocahontas [60] by Marquis [19].
5th d. Bonny [27] by *Jock* [10].
6th d. *White Bag* [5] Imported.

[436]   CURLY.

Calved January 14th, 1882.

Bred by Mr. William Edie, Dunnville, Co. Monck, Ont.; the property of Messrs. McKay & Burleigh, Mechanicsville, Iowa, U.S.

Sire Nelson [435].
Dam Gipsy Queen [194] by Our John [106].
2nd d. Dido [88] by Victor [122].
3rd d. Magnolia [84] by Black Jock [18].
4th d. Galloway Lass [42] by *Jock* [10].
5th *Jet* [2] Imported.

[438]   BLUE BELL.

Calved April 2nd, 1874.

Bred by Mr. J. N. Smith, Bath, Co. Clinton, Mich., U.S.; the property of Mr. Philo Lasher, Coffeysburg, Co. Davies, Mo., U.S.

Sire Duncan [437].
Dam Belle Mahone [196] by Our John [106].
2nd d. Nelly Grey [109] by Black Jock [18].
3rd d. Jessie [49] by Black Jock [18].
4th d. *Maggie* [48] by *Moss Trooper 2nd* [48] (151).

Produce: 1876, Black Baby [451]; 1878, Peter Cope [458]; 1880, Lady Bell [450].

[442]   ONTARIO.

Calved April 15th, 1878.

Bred by and the property of Mr. Daniel Steinmetz, Schoeneck, Pa., U.S.
Sire Wellington Boy [432].
Dam Lady Hamilton [274] by Lord Kenmure [271].
2nd d. Lady Kenmure [140] by *Jock* [10].
3rd d. Pocahontas [60] by Marquis [19].
4th d. Bonny [27] by *Jock* [10].
5th d. *White Bag* [5] Imported.

Produce: 1881, Lucy [434]; 1882, Lady Hope [612].

[444]   SNOWDRIFT.

Calved March 20th, 1874.

Bred by Mr. J. N. Smith, Bath, Co. Clinton, Mich., U.S. ; the property of Mr. Philo Lasher, Coffeysburg, Co. Davies, Mo., U.S.

Sire Napier [346].
Dam Snowball [314] by Victor [122].
2nd d. Blooming Heather 2nd [80] by Dred [15].
3rd d. *Blooming Heather* [53] by *Moss Trooper 2nd* [48] (151).
4th d. *Mary* by *Fergy* (19).
5th d. *Jane* by *Bob*.

Produce: 1876, Scottish Chief [455]; 1877, Black Bessy [453]; 1878, Charley Cope [457]; 1879, Johnny Cope 2nd [456]; 1880, Topsy [454]; 1882, Sir William [619].

[445]   LASSIE.

Calved December 18th, 1876.

Bred by the Nebraska State University, Lincoln, Neb., U.S. ; the property of the Kansas State Agricultural College, Manhattan, Kansas, U.S.

Sire McNeil [443].
Dam Snowflake [315] by Johnny Scott [313].
2nd d. Snowball [314] by Victor [122].
3rd d. Blooming Heather 2nd [80] by Dred [15].
4th d. *Blooming Heather* [53] by *Moss Trooper 2nd* [48] (151).
5th d. *Mary* by *Fergy* (19).
6th d. *Jane* by *Bob*.

Produce: 1879, Jim Crow [430]; 1881, Maggie [452].

[446]   LADY JANE.

Calved March 13th, 1880.

Bred by Mr. Wm. Edie, Dunnville, Co. Monck, Ont. ; the property of Messrs. McKay & Burleigh, Mechanicsville, Iowa, U.S.

Sire Nelson [435].
Dam Gipsy Queen [194] by Our John [106].
2nd d. Dido [88] by Victor [122].
3rd d. Magnolia [84] by Black Jock [18].
4th d. Galloway Lass [42] by *Jock* [10].
5th d. *Jet* [2] Imported.

[448] JENNY LIND.

Calved February 18th, 1877.

Bred by and the property of Mr. Wm. Hood, Guelph, Co. Wellington, Ont.

Sire Black Jock 2nd [253].
Dam Gipsy Queen [194] by Our John [106].
2nd d. Dido [88] by Victor [122].
3rd d. Magnolia [84] by Black Jock [18].
4th d. Galloway Lass [42] by *Jock* [10].
5th d. *Jet* [2] Imported.

Produce: 1880, Lady Black [462]; 1881, Black Jock [448]; 1882, Black Bess [459].

[450] LADY BELL.

Calved March 20th, 1880.

Bred by and the property of Mr. Philo Lasher, Coffeysburg, Co. Davies, Mo., U.S.

Sire Johnny Cope [265].
Dam Blue Bell [438] by Duncan [437].
2nd d. Belle Mahone [196] by Our John [106].
3rd d. Nelly Grey [109] by Black Jock [18].
4th d. Jessie [49] by Black Jock [18].
5th d. *Maggie* [48] by *Moss Trooper 2nd* [48] (151).

Produce: 1881, Johnny Cope 4th [489]; 1882, Bonnie Bell [643].

[451] BLACK BABY.

Calved September 10th, 1876.

Bred by and the property of Mr. Philo Lasher, Coffeysburg, Co. Davies, Mo., U.S.

Sire Johnny Cope [265].
Dam Blue Bell [438] by Duncan [437].
2nd d. Belle Mahone [196] by Our John [106].
3rd d. Nelly Grey [109] by Black Jock [18].
4th d. Jessie [49] by Black Jock [18].
5th d. *Maggie* [48] by *Moss Trooper 2nd* [48] (151).

Produce: 1882, Johnny Cope 5th [618].

452] MAGGIE.

Calved December 21st, 1881.

Bred by and the property of the Kansas State Agricultural College, Manhattan, Kansas, U.S.

Sire McNeil [443].
Dam Lassie [445] by McNeil [443].
2nd d. Snowflake [315] by Johnny Scott [313].
3rd d. Snowball [314] by Victor [122].
4th d. Blooming Heather 2nd [80] by Dred [15].
5th d. *Blooming Heather* [53] by *Moss Tooper 2nd* [48] (151).
6th d. *Mary* by *Fergy* (19).
7th d. *Jane* by *Bob*.

[453] BLACK BESSY.

Calved March 20th, 1877.

Bred by Mr. Philo Lasher, Coffeysburg, Co. Davies, Mo., U.S.; the property of Mr. A. B. Matthews, Kansas City, Mo., U.S.

Sire Johnny Cope [265].
Dam Snowdrift [444] by Napier [346].
2nd d. Snowball [314] by Victor [122].
3rd d. Blooming Heather 2nd [80] by Dred [15].
4th d. *Blooming Heather* [53] by *Moss Trooper 2nd* [48] (151).
5th d. *Mary* by *Fergy* (19).
6th d. *Jane* by *Bob*.

[454] TOPSY.

Calved June 15th, 1880.

Bred by and the property of Mr. Philo Lasher, Coffeysburg, Co. Davies, Mo., U.S.

Sire Johnny Cope [265].
Dam Snowdrift [444] by Napier [346].
2nd d. Snowball [314] by Victor [122].
3rd d. Blooming Heather 2nd [80] by Dred [15].
4th d. *Blooming Heather* [53] by *Moss Trooper 2nd* [48] (151).
5th d. *Mary* by *Fergy* (19).
6th d. *Jane* by *Bob*.

Produce: 1882, Black Jess [620].

[459]   BLACK BESS.

Calved January 3rd, 1882.

Bred by and the property of Mr. Wm. Edie, Dunnville, Co. Monck, Ont.

Sire Lorne [447].
Dam Jenny Lind [448] by Black Jock 2nd [253].
2nd d. Gipsy Queen [194] by Our John [106].
3rd d. Dido [88] by Victor [122].
4th d. Magnolia [84] by Black Jock [18].
5th d. Galloway Lass [42] by *Jock* [10].
6th d. *Jet* [2] Imported.

[460]   KATE.

Calved May 28th, 1881.

Bred by and the property of Mr. Wm. Edie, Dunnville, Co. Monck, Ont.

Sire Lorne [447].
Dam Mollie [461] by Nelson [435].
2nd d. Gipsy Queen [194] by Our John [106].
3rd d. Dido [88] by Victor [122].
4th d. Magnolia [84] by Black Jock [18].
5th d. Galloway Lass [42] by *Jock* [10].
6th d. *Jet* [2] Imported.

[461]   MOLLIE.

Calved March 22nd, 1879.

Bred by and the property of Mr. Wm. Edie, Dunnville, Co. Monck, Ont.

Sire Nelson [435].
Dam Gipsy Queen [194] by Our John [106].
2nd d. Dido [88] by Victor [122].
3rd d. Magnolia [84] by Black Jock [18].
4th d. Galloway Lass [42] by *Jock* [10].
5th d. *Jet* [2] Imported.

Produce: 1881, Kate [460].

[462]   LADY BLACK.

Calved March 13th, 1880.

Bred by and the property of Mr. Wm. Edie, Dunnville, Co. Monck, Ont.

Sire Nelson [435].

Dam Jenny Lind [448] by Black Jock 2nd [253].

2nd d. Gipsy Queen [194] by Our John [106].

3rd d. Dido [88] by Victor [122].

4th d. Magnolia [84] by Black Jock [18].

5th d. Galloway Lass [42] by *Jock* [10].

6th d. *Jet* [2] Imported.

[463]   *PRIMROSE OF KIRKCONNEL* (3317).

Calved 1872.

Imported by Mr. Thos. McCrae, Guelph, Co. Wellington, Ont.; the property of Mr. F. McHardy, Emporia, Kansas, U.S.

Dam *Maggie*, in Scotland, from the herd of Mr. Potts, Craigs, Bewcastle, Brampton, Cumberland, England.

Produce: 1874, Little Emily of Kirkconnel [464]; 1882, The Earl [472].

[464]   *LITTLE EMILY OF KIRKCONNEL* (3321).

Calved 1874.

Imported by Mr. Thomas McCrae, Guelph, Co. Wellington, Ont., from the herd of Mr. J. Graham, Kirkconnel, Castle Douglas, Scotland; the property of Mr. F. McHardy, Emporia, Kansas, U.S.

Sire *Bruce* by *Norman* (529).

Dam *Primrose of Kirkconnel* [463] (3317).

2nd d. *Maggie*, from the herd of Mr. Potts, Bewcastle, Cumberland, England.

[465]   *JEANNIE DEANS SECOND OF TARBREOCH* (3599).

Calved April 15th, 1878.

Imported by Mr. Thos. McCrae, Guelph, Co. Wellington, Ont., from the herd of Messrs. Thos. Biggar & Sons, Chapelton, Dalbeattie, Scotland; the property of Mr. G. S. Burleigh, Mechanicsville, Iowa, U.S.

Sire *Lord of Nithsdale* (616).

Dam *Jeannie Deans* (2696) by *Tom*.

2nd d. *Curly*, from the herd of Mr. John Underwood, Crofts, Kirkcudbright, Scotland.

[466]   *LADY SHANNON.*

Calved 1879.

Imported by Mr. Thos. McCrae, Guelph, Co. Wellington, Ont., from the herd of Mr. Wm. Shannon, Barmoffity, Dalbeattie, Scotland; the property of Mr. G. S. Burleigh, Mechanicsville, Iowa, U.S.

Sire *Sonsy John* (1238).
Dam *Meg of Barmoffity* (3116) by *Flying Dutchman* (547).
2nd d. *Jenny of Barmoffity* (3112) by *Geordie 2nd* (528).
3rd d. *Kate of Balig* (1355) by *Geordie of Rigfoot* (234).
4th d. *Old Kate*, from the herd of Mr. W. J. Shennan, Balig, Kirkcudbright, Scotland.

[467]   *JANE SETON 2nd* (3787).

Calved January 26th, 1879.

Imported by Mr. Thos. McCrae, Guelph, Co. Wellington, Ont., from the herd of Mr. Thos. Biggar & Son, Chapelton, Dalbeattie, Scotland; the property of Mr. F. McHardy, Emporia, Kansas, U.S.

Sire *Lord of Nithsdale* (616).
Dam *Jane Seton* (2697) by *Kirkey*, from the herd of Mr. John Underwood, Crofts, Kirkcudbright, Scotland.

Produce : 1882, Dalbeattie [473].

[468]   *COUNTESS OF GALLOWAY.*

Calved 1879.

Imported by Mr. Thos. McCrae, Guelph, Co. Wellington, Ont., from the herd of the Earl of Galloway, Galloway House, Garliestown, Scotland; the property of Mr. F. McHardy, Emporia, Kansas, U.S.

Sire *Scottish Borderer* (669).
Dam a pure bred Galloway cow from the herd of the Earl of Galloway, Scotland.

Produce : 1882, McLeod of Dare [471].

COWS.

[469] *HELEN OF KIRKCONNEL.*

Calved January 23rd, 1882.

Imported by Mr. Thos. McCrae, Guelph, Co. Wellington, Ont., from the herd of Mr. John Graham, Kirkconnel, Castle Douglas, Scotland; the property of Mr. F. McHardy, Emporia, Kansas, U.S.

Sire *Earlston* [478].
Dam *Little Emily of Kirkconnel* [464] (3320) by *Bruce* by *Norman* (529).
2nd d. *Primrose of Kirkconnel* [463] (3311).
3rd d. *Maggie*, from the herd of Mr. Potts, Bewcastle, Cumberland, England.

[476] MAGGIE SHAW.

Calved July 29th, 1879.

Bred by Mr. Thos. McCrae, Guelph, Co. Wellington, Ont.; the property of Mr. F. McHardy, Emporia, Kansas, U.S.

Sire Lord Lochinvar [405].
Dam Maggie Laidlaw [382] by *Pride of the Speed* (159).
2nd d. Maggie [120] by Dred [15].
3rd d. *Polly Shaw* [174] Imported.

[477] FANNY BLAIR.

Calved May 31st, 1879.

Bred by and the property of Mr. Thomas McCrae, Guelph, Co. Wellington, Ont.

Sire Wellington [404].
Dam Mary Hood [383] by Black Jock 2nd [253].
2nd d. Prairie Lass [261] by Robin [166].
3rd d. Lucy [502] by Black Jock [18].
4th d. Jet [343] by Prince Albert [190].
5th d. *Black Bess* [38] Imported.

Produce: 1882, Nancy Blair [483].

[179]   MEG HOOD.

Calved April 14th, 1874.

Bred by Mr. Wm. Hood, Guelph, Co. Wellington, Ont.; the property of Mr. F. McHardy, Emporia, Kansas, U.S.

Sire Roger [198].
Dam Black Sall [223] by Black Tom [279].
2nd d. White Bag [220] by Moss Trooper 4th [493].
3rd d. Black Swan [219] by Prince Albert [190].
4th d. Black Jessie [218] by *Jock* |10].
5th d. *Sall* [14] Imported.

[482]   FAIRY DALE.

Calved November 15th, 1881.

Bred by Mr. Christopher Wilson, Fergus, Co. Wellington, Ont.; the property of Mr. F. McHardy, Emporia, Kansas, U.S.

Sire Fred 2nd [480].
Dam Lily Dale [233] by Prince Oscar [151].
2nd d. Jennie [152] by Garibaldi [130].
3rd d. Victoria [150] by *Black Prince* [152].
4th d. Beauty [151] by *Peter* (116).

[483]   NANCY BLAIR.

Calved May 5th, 1882.

Bred by Mr. Andrew Elliott, Galt, Co. Waterloo, Ont.; the property of Mr. F. McHardy, Emporia, Kansas, U.S.

Sire Royal Charlie [474].
Dam Fanny Blair [477] by Wellington [404].
2nd d. Mary Hood [383] by Black Jock 2nd [253].
3rd d. Prairie Lass [261] by Robin [166].
4th d. Lucy [502] by Black Jock [18].
5th d. Jet [343] by Prince Albert [190].
6th d. *Black Bess* [6] Imported.

[485]   SALL.

Calved July 1st, 1874.

Bred by and the property of Mr. Wm. Hood, Guelph, Co. Wellington, Ont.

Sire Roger [198].
Dam Lady Isabella [100] by Donald [123].
2nd d. *Chloe* [4] Imported.

Produce: 1878, Fred 2nd [480]; 1881, Fred 4th [481].

[486] QUEEN MARY.

Calved March 16th, 1880.

Bred by Mr. Thos. McCrae, Guelph, Co. Wellington, Ont.; the property of Mr. J. N. Smith, Bath, Mich., U.S.

Sire Lord Douglas [406].
Dam Queen Mab [206] by *Pride of the Speed* [159].
2nd d. May Queen [143] by Honest Tom [98].
3rd d. Queen of Beauty [95] by Black Jock [18].
4th d. *Black Bess* [6] Imported.

[487] POCAHONTAS 3RD.

Calved April 21st, 1878.

Bred by Mr. Thos. McCrae, Guelph, Co. Wellington, Ont.; the property of Mr. E. C. Clinkscales, Columbia, Mo., U.S.

Sire Clandeboye [272].
Dam Galloway Lass [234] by *Pride of the Speed* [159].
2nd d. Maümee [92] by *Jock* [10].
3rd d. *Newbie Lass* [75] Imported.

[488] MISS LAUDER.

Calved May 10th, 1881.

Bred by and the property of Mr. J. N. Smith, Bath, Mich., U.S.

Sire Billy McNeil [327].
Dam Maggie Lauder 2nd [347] by Our John [106].
2nd d. Maggie Lauder [110] by Black Jock [18].
3rd d. Empress [50] by *Jock* [10].
4th d. *Beauty* [11] Imported.

[492] BONNIE BESSIE.

Calved March 4th, 1876.

Bred by Mr. Thos. McCrae, Guelph, Co. Wellington, Ont.; the property of Mr. F. McHardy, Emporia, Kansas, U.S.

Sire Lord Kenmure [271].
Dam Jeannie Deans [243] by King Tom [175].
2nd d. Mary Hay [147] by Dred [15].
3rd d. *Polly Shaw* [174] Imported.

Produce: 1879, Bonnie Lassie [408].

[495]   QUEEN OF THE DAIRY.
Calved June 10th, 1872.
Bred by Mr. Wm. Hood, Guelph, Co. Wellington, Ont.; the property of Mr. Wm. P. Darrow, Jefferson, Co. Hillsdale, Mich., U.S.
Sire Our John [106].
Dam Maggie Lauder [110] by Black Jock [18].
2nd d. Empress [50] by *Jock* [10].
3rd d. *Beauty* [11] Imported.

Produce: 1874, The Nymph [503]; 1879, Black Beauty [500]; 1881, Dark Lady [505].

[496]   QUEEN OF THE BLACKS.
Calved September 10th, 1880.
Bred by Mr. Wm. P. Darrow, Jefferson, Co. Hillsdale, Mich., U.S.; the property of Mr. H. A. Disbrow, Atlantic, Iowa, U.S.
Sire Robin Hood [349].
Dam The Nymph [503] by Shoo-Fly [184].
2nd d. Queen of the Dairy [495] by Our John [106].
3rd d. Maggie Lauder [110] by Black Jock [18].
4th d. Empress [50] by *Jock* [10].
5th d. *Beauty* [11] Imported.

[497]   PRIDE OF THE DAIRY.
Calved June 23rd, 1870.
Bred by Mr. Wm. Hood, Guelph, Co. Wellington, Ont.; the property of Mr. George Coleman, Howell, Co. Livingston, Mich., U.S.
Sire Hardfortune 2nd [255].
Dam Margaret [285] by *Dryfe* [222].
2nd d. Black Bess [38] by Black Tom [37].
3rd d. *Bell* [9] Imported.

Produce: 1873, Herd Laddie [494]; 1874, Chloe [584]; 1878, Robin Hood [501].

[498]   PET.
Calved May 4th, 1878.
Bred by Mr. Jasper Coleman, St. John, Co. Clinton, Mich., U.S.; the property of Mr. Wm. P. Darrow, Jefferson, Co. Hillsdale, Mich., U.S.
Sire Robin Hood [349].
Dam Maggie Lauder [110] by Black Jock [18].
2nd d. Empress [50] by *Jock* [10].
3rd d. *Beauty* [11] Imported.

Produce: 1880, Nigra [499]; 1881, Topsy [506]; 1882, Mink [547].

COWS.

[499] NIGRA.

Calved May 15th, 1880.

Bred by Mr. Wm. P. Darrow, Jefferson, Co. Hillsdale, Mich., U.S.; the property of Mr. H. A. Disbrow, Atlantic, Iowa, U.S.

Sire Hardfortune 2nd [255].
Dam Pet [498] by Robin Hood [349].
2nd d. Maggie Lauder [110] by Black Jock [18].
3rd d. Empress [50] by *Jock* [10].
4th d. *Beauty* [11] Imported.

[500] BLACK BEAUTY.

Calved June 25th, 1879.

Bred by Mr. Wm. P. Darrow, Jefferson, Co. Hillsdale, Mich., U.S.; the property of Mr. H. A. Disbrow, Atlantic, Iowa, U.S.

Sire Robin Hood [349].
Dam Queen of the Dairy [495] by Our John [106].
2nd d. Maggie Lauder [110] by Black Jock [18].
3rd d. Empress [50] by *Jock* [10].
4th d. *Beauty* [11] Imported.

Produce: 1881, Beauty 2nd [507]; 1882, Roderick Dhu [637].

[502] LUCY.

Calved August 2nd, 1870.

Bred by and the property of Mr. Wm. Hood, Guelph, Co. Wellington, Ont.

Sire Black Jock [18].
Dam Jet [343] by Prince Albert [190].
2nd d. *Black Bess* [6] Imported.

Produce: 1871, Prairie Lass [261]; 1872, Robin Hood 2nd [240]; 1874, Wandering Nellie [268].

[503] THE NYMPH.

Calved April 8th, 1874.

Bred by Mr. Jasper Coleman, St. John, Co. Clinton, Mich., U.S.; the property of Mr. Wm. P. Darrow, Jefferson, Co. Hillsdale, Mich., U.S.

Sire Shoo-Fly [184].
Dam Queen of the Dairy [495] by Our John [106].
2nd d. Maggie Lauder [110] by Black Jock [18].
3rd d. Empress [50] by *Jock* [10].
4th d. *Beauty* [11] Imported.

Produce: 1880, Queen of the Blacks [496]; 1881, Meg [656].

[504] JEANNETTE.

Calved March 10th, 1882.

Bred by Mr. Peter Davy, Monterey, Co. Waukesha, Wis., U.S.; the property of Mr. Samuel Martin, Eldora, Co. Hardin, Iowa, U.S.

Sire Johnny Cope [265].
Dam Mary [281] by Bob [252].
2nd d. Fancy [202] by Hardfortune [154].
3rd d. Woolwich Queen [96] by William Wallace [28].
4th d. Black Bess [38] by Black Tom [37].
5th d. *Bell* [9] Imported.

[505] DARK LADY.

Calved March 15th, 1881.

Bred by and the property of Mr. Wm. P. Darrow, Jefferson, Co. Hillsdale, Mich., U.S.

Sire Rob Roy [501].
Dam Queen of the Dairy [495] by Our John [106].
2nd d. Maggie Lauder [110] by Black Jock [18].
3rd d. Empress [50] by *Jock* [10].
4th d. *Beauty* [11] Imported.

[506] TOPSY.

Calved April 10th, 1881.

Bred by and the property of Mr. Wm. P. Darrow, Jefferson, Co. Hillsdale, Mich., U.S.

Sire Rob Roy [501].
Dam Pet [498] by Robin Hood [349].
2nd d. Maggie Lauder [110] by Black Jock [18].
3rd d. Empress [50] by *Jock* [10].
4th d. *Beauty* [11] Imported.

[507] BEAUTY 2ND.

Calved June 1st, 1881.

Bred by and the property of Mr. Wm. P. Darrow, Jefferson, Co. Hillsdale, Mich., U.S.

Sire Rob Roy [501].
Dam Black Beauty [500] by Robin Hood [349].
2nd d. Queen of the Dairy [495] by Our John [106].
3rd d. Maggie Lauder [110] by Black Jock [18].
4th d. Empress [50] by *Jock* [10].
5th d. *Beauty* [11] Imported.

[509] MAGGIE 4TH.

Calved March 10th, 1882.

Bred by and the property of Mr. R. B. Caruss, St. John, Co. Clinton, Mich., U.S.

Sire Lord Lochinvar [405].
Dam Maggie 3rd [364] by Johnny Hood [354].
2nd d. Maggie Lauder 3rd [353] by Robin Hood [349].
3rd d. Maggie Lauder [110] by Black Jock [18].
4th d. Empress [50] by *Jock* [10].
5th d. *Beauty* [11] Imported.

[510] ROSY.

Calved April 10th, 1881.

Bred by and the property of Mr. R. B. Caruss, St. John, Co. Clinton, Mich., U.S.

Sire Johnny Hood [354].
Dam Rosy McNeil [345] by Hardfortune [154].
2nd d. Lizzie [114] by Dred [15].
3rd d. Bonny 2nd [74] by *Jock* [10].
4th d. *Chloe* [4] Imported.

[511] CLINTON.

Calved March 13th, 1882.

Bred by and the property of Mr. R. B. Caruss, St. John, Co. Clinton, Mich., U.S.

Sire Lord Lochinvar [405].
Dam Hannah [512] by Hardfortune [154].
2nd d. Dairymaid [286] by *Dryfe* [222].
3rd d. Black Bess [38] by Black Tom [37].
4th d. *Bell* [9] Imported.

[512] HANNAH.

Calved January, 1871.

Bred by Mr. Floyd Coleman, St. John, Co. Clinton, Mich., U.S.; the property of Mr. R. B. Caruss, St. John, Co. Clinton, Mich., U.S.

Sire Hardfortune [154].
Dam Dairymaid [286] by *Dryfe* [222].
2nd d. Black Bess [38] by Black Tom [37].
3rd d. *Bell* [9] Imported.

Produce: 1881, Hannah 2nd [513]; 1882, Clinton [511].

[513]   HANNAH 2ND.

Calved March 6th, 1881.

Bred by and the property of Mr. R. B. Caruss, St. John, Co. Clinton, Mich., U.S.

Sire Johnny Hood [354].
Dam Hannah [512] by Hardfortune [154].
2nd d. Dairymaid [286] by *Dryfe* [222].
3rd d. Black Bess [38] by Black Tom [37].
4th d. *Bell* [9] Imported.

[514]   POLLY.

Calved December 27th, 1877.

Bred by Mr. Wm. McCrae, Manor Bank, Guelph, Co. Wellington, Ont.; the property of Mr. R. B. Caruss, St. John, Co. Clinton, Mich., U.S.

Sire Major Gray [273].
Dam Mary [111] by Our John [106].
2nd d. Empress [50] by *Jock* [10].
3rd d. *Beauty* [11] Imported.

Produce : 1882, Susie [515].

[515]   SUSIE.

Calved April 23rd, 1882.

Bred by and the property of Mr. R. B. Caruss, St. John, Co. Clinton, Mich., U.S.

Sire Lord Lochinvar [405].
Dam Polly [514] by Major Gray [273].
2nd d. Mary [111] by Our John [106].
2nd d. Empress [50] by *Jock* [10].
3rd d. *Beauty* [11] Imported.

[517]   EMPRESS.

Calved December 9th, 1880.

Bred by and the property of Mr. Joseph Hickson, Montreal, Que.

Sire Laird of Burleigh [516].
Dam Black Swan [328] by Lord Wellington [332].
Dam Betty Laidlaw [276] by Lord Kenmure [271].
2nd d. Maggie Laidlaw [382] by *Pride of the Speed* [159].
3rd d. Maggie [120] by Dred [15].
4th d. *Polly Shaw* [174] Imported.

COWS.

[518]   THE SQUAW.

Calved April 25th, 1880.

Bred by and the property of Mr. Joseph Hickson, Montreal, Que.
Sire Lord Wellington [332].
Dam Betty Laidlaw [276] by Lord Kenmure [271].
2nd d. Maggie Laidlaw [382] by *Pride of the Speed* [159].
3rd d. Maggie [120] by Dred [15].
4th d. *Polly Shaw* [174] Imported.

[522]   BORDER BELLE.

Calved October 23rd, 1878.

Bred by Mr. Wm. McCrae, Manor Bank, Guelph, Co. Wellington, Ont.;
the property of Mr. James Calvert, Dromore, Co. Grey, Ont.
Sire Major Gray [273].
Dam Border Lass [384] by *Young Lochinvar* [303].
2nd d. Minnie [178] by Dred [15].
2rd d. *Heather Bell* [12] Imported.
Produce : 1881, Miss Steele [523]; 1882, Mary [524].

[523]   MISS STEELE.

Calved February 13th, 1881.

Bred by and the property of Mr. James Calvert, Dromore, Co. Grey, Ont.
Sire Lord Chelmsford [521].
Dam Border Belle [522] by Major Gray [273].
2nd d. Border Lass [384] by *Young Lochinvar* [303].
3rd. d. Minnie [178] by Dred [15].
4th d. *Heather Bell* [12] Imported.

[524]   MARY.

Calved January 27th, 1882.

Bred by and the property of Mr. James Calvert, Dromore, Co. Grey, Ont.
Sire Lord Chelmsford [521].
Dam Border Belle [522] by Major Gray [273].
2nd d. Border Lass [384] by *Young Lochinvar* [303].
3rd d. Minnie [178] by Dred [15].
4th d. *Heather Bell* [12] Imported.

[525] *BLACKIE SECOND OF KELLS* (4188).

Calved January 28th, 1881.

Bred by Messrs. Thos. Biggar & Sons, Chapelton, Dalbeattie, Galloway, Scotland; imported by and the property of Mr. Peter Davy, Monterey, Co. Waukesha, Wis., U.S.

Sire *Earl of Nithsdale* (1035).

Dam *Blackie of Kells* (3146), from the herd of Mr. Wm. Sinclair, Trochie House, New Galloway, Scotland.

[526] *CLARET FOURTH* (4187).

Calved February 3rd, 1881.

Bred by Messrs. Thos. Biggar & Sons, Chapelton, Dalbeattie, Galloway, Scotland; imported by and the property of Mr. Peter Davy, Monterey, Co. Waukesha, Wis., U.S.

Sire *Earl of Nithsdale* (1035).

Dam *Claret 3rd* (2135) by *Talisman*, bred by Mr. John Fraser, Knells, Carlisle.

2nd d. *Claret 2nd* (1377) by *Henry* (1127).

3rd d. *Miss Wallace*, from the herd of Mr. John Wallace, Laiybarns, Kirkcudbrightshire, Scotland.

[527] *RANEE THIRD* (4185).

Calved March 21st, 1881.

Bred by Messrs. Thos. Biggar & Sons, Chapelton, Dalbeattie, Galloway, Scotland; imported by and the property of Mr. Peter Davy, Monterey, Co. Waukesha, Wis., U.S.

Sire *Earl of Nithsdale* (1035).

Dam *Ranee* by *Underwood's Bull*.

2nd d. *Miss Good* by *Cunningham's Bull*.

[528] *LADY KEN*.

Calved March 3rd, 1880.

Bred by Messrs. Thos. Biggar & Sons, Chapelton, Dalbeattie, Galloway, Scotland; imported by and the property of Mr. Peter Davy, Monterey, Co. Waukesha, Wis., U.S.

Sire *Earl of Nithsdale* (1035).

Dam *McQueen*, a pure Galloway cow.

COWS.

[531] *MOSS ROSE THIRD OF WEDHOLME* (3562).

Calved April 11th, 1878.

Bred by Mr. John Millican, Wedholme House, Abbey Town, Scotland imported by and the property of Mr. Peter Davy, Monterey, Co. Waukesha, Wis., U.S.

Sire *Hazeldean* (1010).
Dam *Moss Rose of Wedholme* (2876) by *Peddar Lad* (1139).

[532] *MOSS ROSE FOURTH* (3721).

Calved April 2nd, 1879.

Bred by Mr. John Millican, Wedholme House, Abbey Town, Scotland; imported by and the property of Mr. Peter Davy, Monterey, Co. Waukesha, Wis., U.S.

Sire *Hazeldean* (1010).
Dam *Moss Rose 2nd of Wedholme* (2881) by *Black Jock of Pelutho* (576).
2nd d. *Moss Rose of Wedholme* (2876) by *Peddar Lad* (1139).

[533] *ROSE FIFTH OF WEDHOLME* (3948).

Calved January 7th, 1880.

Bred by Mr. John Millican, Wedholme House, Abbey Town, Scotland; imported by and the property of Mr. Peter Davy, Monterey, Co. Waukesha, Wis., U.S.

Sire *The Mackintosh 2nd* (1341).
Dam *Rose 4th of Wedholme* (3183) by *Lincoln 2nd* (1001).
2nd d. *Rose 2nd of Wedholme* (2880) by *Peddar Lad* (1139).

[534] *SEA BIRD.*

Calved June 18th, 1882.

Bred by Messrs. Thos. Biggar & Sons, Chapelton, Dalbeattie, Galloway, Scotland; imported by and the property of Mr. Peter Davy, Monterey, Co. Waukesha, Wis., U.S.

Sire *Investment* (1578).
Dam *Lady Ken* [528] by *Earl of Nithsdale* (1035).
2nd d. *McQueen*, a pure Galloway cow.

[535] ALICE MAUD (3977).

Calved February 2nd, 1879.

Bred by Mr. John Carruthers, Kirkhill, Moffat, Scotland; imported by and the property of Mr. Peter Davy, Monterey, Co. Waukesha, Wis., U.S.

Sire *Samson* (1179).
Dam *Queen of Kirkhill* (3331) by *Moorcock of Kirkhill* (1308).
2nd d. *Lady* by *Donald*, bred by Mr. Keir, of Whithaugh.
3rd d. *Bloomer* by *Blackesk*.
4th d. *Victoria* by *Ranger*.
5th d. *Rosebud* by *Black Jock of Riggfoot* (66).

[536] BELL.

Calved October, 1880.

Bred by Mr. Francis Carruthers, Cleughheads, Lockerbie, Scotland; imported by and the property of Mr. Peter Davy, Monterey, Co. Waukesha, Wis., U.S.

Sire *Willie Hope of Balgray* (1216).
Dam a pure Galloway cow, bred by Mr. Francis Carruthers

[537] DAMSON SECOND.

Calved April, 1880.

Bred by Mr. Maxwell, Castle Douglas, Scotland; imported by and the property of Mr. Peter Davy, Monterey, Co. Waukesha, Wis., U.S.

Sire *Huntsman of Culmain* (1048).
Dam *Damson of Screel* (3348), from the herd of Mr. Adam Thomson, Screel, Castle Douglas, Scotland.

[538] KATE.

Calved January 26th, 1880.

Bred by Mr. James Cunningham, Tarbreoch, Dalbeattie, Scotland; imported by and the property of Mr. Peter Davy, Monterey, Co. Waukesha, Wis., U.S.

Sire *Knowsley* (1279).
Dam *Beauty of Talton Wheat* (2767) by *Bob of the Bent* (1125).
2nd d. *Dolly 2nd* (2763).
3rd d. *Dolly 1st*, from the herd of the Rev. Geo. Murray, Balmaclellan Manse, New Galloway.

[539] *NELLIE.*

Calved May, 1880.

Bred by Mr. Francis Carruthers, Cleughheads, Lockerbie, Scotland; imported by and the property of Mr. Peter Davy, Monterey, Co. Waukesha, Wis., U.S.

Sire *Willie Hope of Balgray* (1216).
Dam a pure Galloway cow, bred by Mr. Francis Carruthers.

[540] *PRINCESS MARY.*

Calved February 10th, 1880.

Bred by Mr. Hugh Josselyn Percy, Eskrigg House, Scotland; imported by and the property of Mr. Peter Davy, Monterey, Co. Waukesha, Wis., U.S.

Sire *Prince Victor* (1473).
Dam *Primrose of Pelutho* (1368) by *Camerton* (542).
2nd d. *Daisy*, bred by Mr. Thos. Chambers, Pelutho, Abbey Town, Carlisle.

[541] *QUEEN OF ESKRIGG.*

Calved February 20th, 1880.

Bred by Mr. Hugh Josselyn Percy, Eskrigg, House, Scotland; imported by and the property of Mr. Peter Davy, Monterey, Co. Waukesha, Wis., U.S.

Sire *Prince Victor* (1473).
Dam *Beauty of Pelutho 2nd* (1365) by *Camerton* (542).
2nd d. *Beauty of Pelutho* (1358) by *Geordie 2nd of Green* (1135).
3rd d. *Whitrigg*, from the herd of Mr. Thos. Chambers, Pelutho, Abbey Town, Carlisle.

[542] *SLOE SECOND.*

Calved March, 1880.

Bred by Mr. Maxwell, Castle Douglas, Scotland; imported by and the property of Mr. Peter Davy, Monterey, Co. Waukesha, Wis., U.S.

Sire *Huntsman of Culmain* (1048).
Dam *Sloe of Screel* (3346), from the herd of Mr. Adam Thomson, Screel, Castle Douglas, Scotland.

[544] CURLY NELL.

Calved August 31st, 1882.

Bred by and the property of Messrs. David Corey & Son, Yates, Co. Knox, Ill., U.S.

Sire Johnny Cope [265].
Dam Jessie 2nd [320] by Sam [280].
2nd d. Jessie [267] by Black Tom [279].
3rd d. Wandering Nellie [277] by Wm. Wallace [28].
4th d. Miss Torrence [204] by Wm. Wallace [28].
5th d. Black Jane [210] by Jacob Benner's Bull [196].
6th d. *Black Bess* [6] Imported.

[545] MAUD.

Calved April 25th, 1882.

Bred by and the property of Mr. Peter Davy, Monterey, Co. Waukesha, Wis., U.S.

Sire Johnny Cope [265].
Dam Rosa [326] by Sam [280].
2nd d. Dora Bell [297] by Wellington Boy [264].
3rd d. Sis [203] by Hardfortune [154].
4th d. Susan [157] by *Dryfe* [222].
5th d. Bess [125] by Marquis [19].
6th d. *Chloe* [4] Imported.

[547] MINK.

Calved April 15th, 1882.

Bred by Mr. Wm. P. Darrow, Jefferson, Co. Hillsdale, Mich., U.S.; the property of Mr. Z. R. Ashbaugh, Hillsdale, Co. Hillsdale, Mich., U.S.

Sire Rob Roy [501].
Dam Pet [498] by Robin Hood [349].
2nd d. Maggie Lauder [110] by Black Jock [18].
3rd d. Empress [50] by *Jock* [10].
4th d. *Beauty* [11] Imported.

COWS.

[556]   *BEAUTY SECOND OF TROQUHAIN* (3517).

Calved February 15th, 1878.

Bred by Messrs. M. & J. S. Wilson, New Galloway, Scotland; imported by and the property of Mr. Thomas McCrae, Guelph, Co. Wellington, Ont.

Sire *Watty* (1072).

Dam *Beauty of Troquhain* (2806) by *Jugurtha*.

2nd d. *Missie*, from the herd of Mr. Wilson, Troquhain, New Galloway, Scotland.

[557]   *BELTED LASS.*

Calved 1878.

Bred by Mr. Thos. Fisher, Craignorget, Wigtonshire, Scotland; imported by and the property of Mr. Thos. McCrae, Guelph, Co. Wellington, Ont.

Sire *Craignorget* [611].

Dam *Mary*, from the herd of Mr. Thos. Fisher, Craignorget.

[558]   *BLOSSOM OF TARBREOCH.*

Calved February 28th, 1880.

Bred by the Earl of Galloway, Garliestown, Scotland; imported by and the property of Mr. Thos. McCrae, Guelph, Co. Wellington, Ont.

Sire *Scottish Borderer* (669).

Dam *Blossom of Garlieston* (2846).

2nd d. *Susie*, from the herd of Mr. John Underwood, Crofts, Castle Douglas, Scotland.

[559]   *CAREFUL.*

Calved 1880.

Bred by Mr. Wm. McMicken, Gatehouse of Fleet, Scotland; imported by and the property of Mr. Thos. McCrae, Guelph, Co. Wellington, Ont.

Sire *Forest King 2nd* (1153).

Dam *White Belly* by *Old Mulock*.

2nd d. from the herd of Mr. John Halliday, Mulock, Kirkcudbrightshire, Scotland.

560] *CARNATION.*

Calved March, 1880.

Bred by Mr. Armstrong, Foster Meadow, Lockerbie, Scotland; imported by and the property of Mr. Thos. McCrae, Guelph, Co. Wellington, Ont.

Sire *Uncle Ned* (1182).
Dam *Jean*, from the herd of Mr. Armstrong.

[561] *CHEERFUL.*

Calved April, 1880.

Bred by Mr. Wm. McMicken, Gatehouse of Fleet, Scotland; imported by and the property of Mr. Thos. McCrae, Guelph, Co. Wellington, Ont.

Sire *Forest King 2nd* (1153).
Dam *Lagghead*, from the herd of Mr. Kelly, Lagghead, Girthon, Kirkcudbrightshire, Scotland.

[562] *COILA.*

Calved February, 1878.

Bred by Mr. Wm. McMicken, Gatehouse of Fleet, Scotland; imported by and the property of Mr. Thos. McCrae, Guelph, Co. Wellington, Ont.

Sire *Mulock Bob 2nd.*
Dam *Jean of Creochs* by *Lairdlaugh.*

[563] *COMFORT.*

Calved 1880.

Bred by Mrs. Fraser, Glaisters, Kirkpatrick, Dereham, Scotland; imported by and the property of Mr. Thos. McCrae, Guelph, Co. Wellington, Ont.

Sire *Macdonald of Lochenkit.*
Dam a pure bred Galloway cow, from the herd of Mrs. Fraser, Glaisters.

COWS.

[564] *CRICKET.*

Calved 1880.

Bred by Mr. Robt. Rae, Meikle Cocklick, Dalbeattie, Scotland; imported by Mr. Thos. McCrae, Guelph, Co. Wellington, Ont.; and the property of Mr. J. N. Smith, Bath, Co. Clinton, Mich., U.S.

Sire *Noble Boy* (1184).

Dam *Jean* by *Oliver* (1102).

2nd d. from the herd of Mr. Robert Rae.

[565] *EDITH OF TROQUHAIN* (3518).

Calved February 5th, 1878.

Bred by Messrs. M. & J. S. Wilson, Troquhain, New Galloway, Scotland; imported by and the property of Mr. Thos. McCrae, Guelph, Co. Wellington, Ont.

Sire *Watty* (1072).

Dam *Nelly of Troquhain* (3128) by *Glaisters.*

2nd d. a pure cow, from the herd of Mr. Alex. McTurk, Borlae, Dalry, New Galloway, Scotland.

[566] *JANE SETON THIRD.*

Calved August 1st, 1880.

Bred by Messrs. Thos. Biggar & Sons, Chapelton, Dalbeattie, Scotland; imported by Mr. Thos. McCrae, Guelph, Co. Wellington, Ont.; the property of Mr. R. B. Caruss, St. John, Co. Clinton, Mich., U.S.

Sire *Earl of Nithsdale* (1035).

Dam *Jane Seton 2nd* (3787) by *Lord of Nithsdale* (616).

2nd d. *Jane Seton* (2697) by *Kirkey*, from the herd of Mr. John Underwood, Crofts, Kirkcudbright, Scotland.

[567] *JENNY OF TARBREOCH.*

Calved March, 1879.

Bred by Mr. Wm. Shennan, Barmoffity, Dalbeattie, Scotland; imported by and the property of Mr. Thos. McCrae, Guelph, Co. Wellington, Ont.

Sire *Sonsy John* (1238).

Dam *Jenny 2nd of Barmoffity* (3114) by *Barmoffity* (1236).

2nd d. *Jenny of Barmoffity* (3112) by *Geordie 2nd* (528).

3rd d. *Kate of Balig* (1355) by *Geordie of Riggjoot* (234).

4th d. *Old Kate*, from the herd of Messrs. W. & J. Shennan, Balig, Kirkcudbright, Scotland.

[568] *LADY GEILLS.*

Calved October 6th, 1882.

Bred by Mr. Thos. Fisher, Craignorget, Wigtonshire, Scotland; imported by and the property of Mr. Thos. McCrae, Guelph, Co. Wellington, Ont.

Sire *Islesman* (1590).
Dam *Belted Lass* [557] by *Craignorget* [611].
2nd d. *Mary*, from the herd of Mr. Thos. Fisher.

[569] *LUCETTA* (3441).

Calved March, 1878.

Bred by the Earl of Galloway, Garliestown, Scotland; imported by and the property of Mr. Thos. McCrae, Guelph, Co. Wellington, Ont.

Sire *Scottish Borderer* (669).
Dam from the herd of Mr. John Cunningham, Whitecairn, Dalbeattie, Scotland.

[570] *LUCETTA SECOND.*

Calved January, 1881.

Bred by Mr. James Cunningham, Tarbreoch, Scotland; imported by and the property of Mr. Thos. McCrae, Guelph, Co. Wellington, Ont.

Sire *Knowsley* (1279).
Dam *Lucetta* [569] (3441) by *Scottish Borderer* (669).
2nd d. from the herd of Mr. John Cunningham.

[571] *MAGGIE SECOND OF KILLIMINGAN* (3877).

Calved January 2nd, 1878.

Bred by Mr. Joseph Neilson, Killimingan, Dumfries, Scotland; imported by and the property of Mr. Thos. McCrae, Guelph, Co. Wellington, Ont.

Sire *Scottish Chief* (1040).
Dam *Maggie of Killimingan* (2745) by *Marksman 3rd* (1245).
2nd d. from the herd of Mrs. Morrison, Lochill, New Abbey, Dumfries, Scotland.

[572] *MAGGIE FOURTH OF ELRIG* (3883).

Calved February 25th, 1880.

Bred by Mr. Wm. Routledge, Port William, Elrig, Scotland; imported by and the property of Mr. Thos. McCrae, Guelph, Co. Wellington, Ont.

Sire *The Baron of Drumlanrig* (1158).
Dam *Maggie 2nd of Elrig* (3030) by *Scottish Borderer* (669).
2nd d. *Maggie of Elrig* (2701) by *Heir-at-Law* (815).
3rd. d. *Jean*, from the herd of Mr. Wm Routledge.

[573] *MAID OF DUMFRIES* (4974).

Calved 1878.

Imported by and the property of Mr. Thos. McCrae, Guelph, Co. Wellington, Ont., from the herd of Mr. John Ferguson, Kilquhanity, Kirkpatrick, Scotland.

Sire *Rob Roy of Chapelton* (1856).
Dam a pure Galloway cow, from the herd of Mr. Ferguson.

[574] *MEG OF CONGEITH.*

Calved March 28th, 1879.

Bred by Mr. John Cannon, Congeith, Kirkgunzeon, Scotland; imported by and the property of Mr. Thos. McCrae, Guelph, Co. Wellington, Ont.

Sire *Lofty* (1085).
Dam *Maggie of Congeith* (3086) by *Scotchman* (1065).
2nd d. from the herd of Mr. James Nish, Cowens, Kirkgunzeon, Dumfries, Scotland.

[575] *MISS HOPE SECOND OF BALGRAY.*

Calved February 13th, 1880.

Bred by Mr. J. Jardine Paterson, Balgray, Lockerbie, Scotland; imported by and the property of Mr. Thos. McCrae, Guelph, Co. Wellington, Ont.

Sire *Olden Times* (1369).
Dam *Miss Hope of Balgray* (3077) by *Bob of Balgray* (1092).
2nd d. *Jenny of Balgray* (2908) by *Othello* (835).
3rd d. from the herd of Mr. J. Jardine Paterson.

[576] *MISS NELLIE OF BELCATHILL.*

Calved March, 1880.

Bred by Mr. John Carruthers, Belcathill, Lockerbie, Scotland; imported by Mr. Thomas McCrae, Guelph, Co. Wellington, Ont.; and the property of Mr. J. N. Smith, Bath, Mich., U. S.

Sire *Coggie of Belcathill.*

Dam a pure cow, from the herd of Mr. Carruthers.

[577] *MISS OF BELCATHILL.*

Calved March, 1880.

Bred by Mr. John Carruthers, Belcathill, Lockerbie, Scotland; imported by Mr. Thos. McCrae, Guelph, Co. Wellington, Ont.; the property of Mr. Mr. R. B. Caruss, St. John, Mich., U.S.

Sire *Coggie of Belcathill.*

Dam a pure cow, from the herd of Mr. Carruthers.

[578] *NANCY SECOND OF BALGRAY.*

Calved March 24th, 1881.

Bred by Mr. J. Jardine Paterson, Balgray, Lockerbie, Scotland; imported by and the property of Mr. Thos. McCrae, Guelph, Co, Wellington, Ont.

Sire *Olden Times* (1369).

Dam *Nancy of Oakbank* (2677) by *Forest King* (553).

2nd d. *Maid of Arden* (1509) by *Glenorchy* (521).

3rd d. *Margaret*, from herd of Geo. Graham, Oakbank, Longtown. Scotland.

[579] *NANCY LEE SECOND.*

Calved February 25th, 1881.

Bred by Messrs. Thomas Biggar & Sons, Chapelton, Dalbeattie, Scotland; imported by and the property of Mr. Thomas McCrae, Guelph, Co. Wellington, Ont.

Sire *Earl of Nithsdale* (1035).

Dam *Nancy of Knocklae* (2874) by *Neil Gow* (1138).

2nd d. *Soncie of Drumlanrig* (2869).

3rd d. *Lily of Knocklae* (2867) by *Blucher* (1136).

4th d. *Old Lily*, from the herd of Mr. Thos. Corrie, Balmaclellan, New Galloway, Scotland.

COWS.

[580] *NANCY OF TARBREOCH.*

Calved January, 1881.

Bred by Mr. James Cunningham, Tarbreoch, Scotland; imported by and the property of Mr. Thomas McCrae, Guelph, Co. Wellington, Ont.

Sire *Knowsley* (1279).
Dam *Mary 5th of Tarbreoch* (3435) by *Chieftain of Drumlanrig* (752).
2nd d. *Bridesmaid of Tarbreoch* (1674) by *Observer* (728).
3rd d. *Mary 2nd of Tarbreoch* (1671) by *Balig* (729).
4th d. from the herd of Mr. Joseph Kerr, Flats of Cargen, Dumfries, Scotland.

[581] *NELLIE OF CORWALL* (3885).

Calved January 5th, 1879.

Bred by Mr. Andrew Milligan, Port William, Scotland; imported by and the property of Mr. Thos. McCrae, Guelph, Co. Wellington, Ont.

Sire *Dominie Sampson* (1149).
Dam *Mary 3rd of Redcastle* (2938) by *Baron Douglas* (614).
2nd d. *Mary of Redcastle* (2926).
4th d. from the herd of Mr. Jas. Graham, Meikle Culloch, Dalbeattie, Scotland.

[582] *TOPSY OF TARBREOCH.*

Calved March 25th, 1880.

Bred by Earl of Galloway, Garliestown, Scotland; imported by and the property of Mr. Thos. McCrae, Guelph, Co. Wellington, Ont.

Sire *Scottish Borderer* (669).
Dam *Topsy of Garlieston* (2843).
2nd d. from the herd of Mr. John McWilliam, Bavernochan, Kirkinner, Scotland.

[583] MOLLY.

Calved April 3rd, 1876.

Bred by Mr. George Coleman, Howell, Co. Livingston, Mich., U.S.; the property of Mr. Wm. P. Darrow, Jefferson, Co. Hillsdale, Mich., U.S.

Sire Napier [346].
Dam Chloe [584] by Herd Laddie [494].
2nd d. Pride of the Dairy [497] by Hardfortune 2nd [255].
3rd d. Margaret [285] by *Dryfe* [222].
4th d. Black Bess [38] by Black Tom [37].
5th d. *Bell* [9] Imported.

[584]   CHLOE.

Calved May 13th, 1874.

Bred by and the property of Mr. George Coleman, Howell, Co. Livingston, Mich., U.S.

Sire Herd Laddie [494].
Dam Pride of the Dairy [497] by Hardfortune 2nd [255].
2nd d. Margaret [285] by *Dryfe* [222].
3rd d. Black Bess [38] by Black Tom [37].
4th d. *Bell* [9] Imported.

Produce: 1876, Molly [583]; 1877, Naomi [619]; 1878, Dinah [613]; 1879, Monah [641]; 1880, Tom [587]; 1882, Rolly Polly [616].

[585]   SWEET AFTON.

Calved April 20th, 1877.

Bred by Mr. George Coleman, Howell, Co. Livingston, Mich., U.S.; the property of Mr. Wm. Keith, Pittsford, Co. Hillsdale, Mich., U.S.

Sire Robin Hood [349].
Dam Chloe [584] by Herd Laddie [494].
2nd d. Pride of the Dairy [497] by Hardfortune 2nd [255].
3rd d. Margaret [285] by *Dryfe* [222].
4th d. Black Bess [38] by Black Tom [37].
5th d. *Bell* [9] Imported.

[586]   DAISY.

Calved December 25th, 1881.

Bred by Mr. Geo. Coleman, Howell, Co. Livingston, Mich., U.S.; the property of Mr. Wm. Keith, Pittsford, Co. Hillsdale, Mich., U.S.

Sire Rob Roy [501].
Dam Sweet Afton (585) by Robin Hood [349].
2nd d. Chloe [584] by Herd Laddie [494].
3rd d. Pride of the Dairy [497] by Hardfortune 2nd [255].
4th d. Margaret [285] by *Dryfe* [222].
5th d. Black Bess [38] by Black Tom [37].
6th d. *Bell* [9] Imported.

[596] *ALICE MAXWELL.*

Calved March, 1875.

Bred by Mr. Adam Thomson, Castle Douglas, Scotland; imported by and the property of Mr. Simon Beattie, Annan, Scotland.

Sire *Tom of Crofts.*

Dam a cow bred by Mr. Adam Thomson, and descended from pure bred Galloway stock.

Produce : 1882, Peter [593].

[597] *BEAUTY.*

Calved May, 1880.

Bred by Mr. Jas. Cunningham, Tarbreoch, Scotland; imported by and the property of Mr. Peter Davy, Monterey, Co. Waukesha, Wis., U. S.

Sire *Knowsley* (1279).

Dam a pure cow, bred by Lord Galloway.

[598] *BELL.*

Calved May, 1880.

Bred by Mr. J. Broach, Waterbeck, Dumfriesshire, Scotland ; imported by and the property of Mr. Simon Beattie, Annan, Scotland.

Sire *Annandale* (1773).

Dam a pure bred Galloway cow.

[599] *BELL OF CLEUGHHEADS.*

Calved April, 1880.

Bred by Mr. Francis Carruthers, Lockerbie, Scotland; imported by and the property of Mr. Simon Beattie, Annan, Scotland.

Sire *Willie Hope of Balgray* (1216).

Dam a cow bred by Mr. F. Carruthers, from pure Galloway stock.

[600] *BLANCHE.*

Calved May, 1880.

Bred by Mr. Francis Carruthers, Lockerbie, Scotland; imported by and the property of Simon Beattie, Annan, Scotland.

Sire *Willie Hope of Balgray* (1216).

Dam a cow bred by Mr. F. Carruthers, from pure Galloway stock.

[601] *BLOOM.*

Calved April, 1881.

Bred by Mr. J. Broach, Waterbeck, Scotland; imported by and the property of Mr. Simon Beattie, Annan, Scotland.

Sire *Annandale* (1773).
Dam a pure bred Galloway cow.

[602] *CHERRY BLOSSOM.*

Calved 1882.

Bred by Mr. Robert Wallace, Kirkcudbright, Scotland; imported by and the property of Mr. Simon Beattie, Annan, Scotland.

Sire *Scotia of Tarbreoch* [594].
Dam *Cherry of Hensol* [603] (3554) by *The Major of Hensol* (1299).
2nd d. *Bell of Hensol* (3304).
3rd d. from the herd of Mr. T. R. Bruce, Slogarie, Balmaghie, Kirkcudbrightshire, Scotland.

[603] *CHERRY OF HENSOL* (3554).

Calved March 6th, 1878.

Bred by Mr. R. De Barre, Cunningham, New Galloway Station, Scotland; imported by and the property of Mr. Simon Beattie, Annan, Scotland.

Sire *The Major of Hensol* (1299).
Dam *Bell of Hensol* (3304).
2nd d. from the herd of Mr. T. R. Bruce.
Produce: 1881, Young Cherry [609]; 1882, Cherry Blossom [602].

[604] *LADY KENMURE.*

Calved 1878.

Bred by Mr. Robert McFadyean, New Galloway, Scotland; imported by and the property of Mr. Simon Beattie, Annan, Scotland.

Sire *Mac* (1852).
Dam *Blackie of Kells* (3146).
2nd d. from the herd of Mr. Wm. Sinclair, Trochie House, New Galloway, Scotland.

COWS.

[605] *MAGGIE.*

Calved July, 1880.

Bred by Mr. Wm. Steel, Short Riggs, Dumfriesshire, Scotland; imported by and the property of Mr. Simon Beattie, Annan, Scotland.

Sire *Olden Times* (1369).

Dam a pure bred Galloway cow, from the herd of Mr. Carruthers, Cleughheads, Lockerbie, Scotland.

[606] *MISS CARRUTHERS.*

Calved April, 1880.

Bred by Mr. Francis Carruthers, Lockerbie, Scotland; imported by and the property of Mr. Simon Beattie, Annan, Scotland.

Sire *Willie Hope of Balgray* (1216).

Dam a cow bred by Mr. F. Carruthers, from pure Galloway stock.

[607] *NANCY.*

Calved June, 1881.

Bred by Mr. Wm. Steel, Short Riggs, Dumfriesshire, Scotland; imported by and the property of Mr. Simon Beattie, Annan, Scotland.

Sire *Olden Times* (1369).

Dam a cow bred by Mr. F. Carruthers, from pure Galloway stock.

[608] *NELLIE.*

Calved April, 1882.

Bred by Mr. Jas. McLean, Annan, Scotland; imported by and the property of Mr. Simon Beattie, Annan, Scotland.

Sire *Maori Chief* [592] (1433).

Dam a pure Galloway cow, from Lord Galloway's stock.

[609] *YOUNG CHERRY.*

Calved 1881.

Bred by Mr. Robert Wallace, Kirkcudbright, Scotland; imported by and the property of Mr. Simon Beattie, Annan, Scotland.

Sire *Major of Buckonhill* (1444).

Dam *Cherry of Hensol* [603] (3554) by *The Major of Hensol* (1299).

2nd d. *Bell of Hensol* (3304).

3rd d. pure cow, from the herd of Mr. T. R. Bruce, Slogarvie, Balmaghie, Kirkcudbrightshire, Scotland.

[610]   *NETTIE*,

Calved May, 1882.

Bred by Mr. Jas. McLean, Annan, Scotland; imported by and the property of Mr. Simon Beattie, Annan, Scotland.
Sire *Maori Chief* [592] (1433).
Dam a pure cow, from Lord Galloway's stock.

[611]   *CRAIGNORGET*.

Calved 1876.

Bred by the Earl of Galloway, Garlieston, Scotland; the property of Mr. Thos. Fisher, Craignorget, Wigtonshire, Scotland.
Sire *Minnigaff* (1076).
Dam a pure cow, bred by the Earl of Galloway.

[612]   LADY HOPE.

Calved April 15th, 1872.

Bred by and the property of Mr. Daniel Steinmetz, Schoeneck, Penn., U.S.
Sire Hopewell [441].
Dam Ontario [442] by Wellington Boy [432].
2nd d. Lady Hamilton [274] by Lord Kenmure [271].
3rd d. Lady Kenmure [140] by *Jock* [10].
4th d. Pocahontas [60] by Marquis [19].
5th d. Bonny [27] by Black Jock [18].
6th d. *White Bag* [5] Imported.

[613]   DINAH.

Calved May 4th, 1878.

Bred by Mr. George Coleman, Howell, Co. Livingston, Mich., U.S.; the property of Mr. J. J. Bush, Lansing, Mich., U.S.
Sire Robin Hood [349].
Dam Chloe [584] by Herd Laddie [494].
2nd d. Pride of the Dairy [497] by Hardfortune 2nd [255].
3rd d. Margaret [285] by *Dryfe* [222].
4th d. Black Bess [38] by Black Tom [37].
5th d. *Bell* [9] Imported.
Produce: 1880, Lord Malcolm [614]: 1882, Garfield [634].

COWS.

[614]   MARION.

Calved April 5th, 1879.

Bred by Mr. Geo. Coleman, Howell, Co. Livingston, Mich., U.S.; the property of Mr. Samuel A. Browne, Pentwater, Co. Oceana, Mich., U. S.

                 Sire Rob Roy [501].
Dam Pride of the Dairy [497] by Hardfortune 2nd [255].
    2nd d. Margaret [285] by *Dryfe* [222].
    3rd. d. Black Bess [38] by Black Tom [37].
      4th d. *Bell* [9] Imported.

Produce: 1881, Dora Deane [615]: 1882, Capitola [647].

[615]   DORA DEANE.

Calved October, 1881.

Bred by Mr. George Coleman, Howell, Co. Livingston, Mich. U. S.; the property of Mr. S. A. Tubbs, Delhi, Co. Washtenaw, Mich., U.S.

          Sire Lord Malcolm [614].
    Dam Marion [614] by Rob Roy [501].
2nd d. Pride of the Dairy [497] by Hardfortune 2nd [255].
    3rd d. Margaret [285] by *Dryfe* [222].
    4th d. Black Bess [38] by Black Tom [37].
      5th d. *Bell* [9] Imported.

[616]   MAID OF LIVINGSTON.

Calved April, 1876.

Bred by and the property of Mr. George Coleman, Howell, Co. Livingston, Mich., U. S.

              Sire Napier [346].
Dam Pride of the Dairy [497] by Hardfortune 2nd [255].
    2nd d. Margaret [285] by *Dryfe* [222].
    3rd d. Black Bess [38] by Black Tom [37].
      4th d. *Bell* [9] Imported.

[617]   SWEET ALICE.

Calved April 15th, 1875.

Bred by and the property of Mr. George Coleman, Howell, Co. Livingston, Mich., U. S.

              Sire Napier [346].
Dam Pride of the Dairy [497] by Hardfortune 2nd [255].
    2nd d. Margaret [285] by *Dryfe* [222].
    3rd d. Black Bess [38] by Black Tom [37].
      4th d. *Bell* [9] Imported.

[618]   BESSIE.

Calved August 1st, 1882.

Bred by and the property of Mr. George Coleman, Howell, Co. Livingston, Michigan, U. S.

Sire Lord Malcolm [614].
Dam Maid of Livingston [616] by Napier [346].
2nd d. Pride of the Dairy [497] by Hardfortune 2nd [255].
3rd d. Margaret [285] by *Dryfe* [222].
4th d. Black Bess [38] by Black Tom [37].
5th d. *Bell* [9] Imported.

[619]   NAOMI.

Calved April 3rd, 1877.

Bred by Mr. George Coleman, Howell, Co. Livingston, Mich., U. S.; the property of Mr. Charles C. Wood, Lamoille, Co. Bureau, Ill., U. S.

Sire Napier [346].
Dam Chloe [584] by Herd Laddie [494].
2nd d. Pride of the Dairy [497] by Hardfortune 2nd [255].
3rd d. Margaret [285] by *Dryfe* [222].
4th d. Black Bess [38] by Black Tom [37].
5th d. *Bell* [9] imported.

Produce: 1882, Lord Malcolm 2nd [635].

[620]   BLACK JESS.

Calved June 15th, 1882.

Bred by and the property of Mr. Philo Lasher, Coffeysburg, Co. Davies, Mo., U. S.

Sire Sir William [340].
Dam Topsy [454] by Johnny Cope [265].
2nd d. Snowdrift [444] by Napier [346].
3rd d. Snowball [314] by Victor [122].
4th d. Blooming Heather 2nd [80] by Dred [15].
5th d. *Blooming Heather* [53] by *Moss Trooper 2nd* [48] (151).
6th d. *Mary* by *Fergy* (19).
7th d. *Jane* by *Bob*.

COWS.

[621] *BLOSSOM.*

Calved May, 1880.

Imported in 1882 by and the property of Mr. Peter Davy, Monterey, Co. Waukesha, Wis., U.S.

Sire *Lariston* (1030).

Dam pure Galloway cow, from the herd of Mr. James Cunningham, Tarbreoch, Castle Douglas, Scotland.

[622] JESSAMINE.

Calved March 14th, 1882.

Bred by Mr. J. J. Rogers, Abingdon, Ill., U.S.; the property of Mr. Ashly Hamilton, Butler, Co. Bates, Mo., U.S.

Sire John Borland [620].

Dam Snip [317] by Billy McNeil [327].

2nd d. Snowflake [315] by Johnny Scott [313].

3rd d. Snowball [314] by Victor [122].

4th d. Blooming Heather 2nd [80] by Dred [15].

5th d. *Blooming Heather* [53] by *Moss Trooper 2nd* [48] (151).

6th d. *Mary* by *Fergy* (19).

7th d. *Jane* by *Bob.*

[623] SNIP 3RD.

Calved January 13th, 1881.

Bred by Mr. Peter Davy, Monterey, Co. Waukesha, Wis., U.S.; the property of Mr. Ashly Hamilton, Butler, Co. Bates, Mo., U.S.

Sire Johnny Scott [313].

Dam Snip [317] by Billy McNeil [327].

2nd d. Snowflake [315] by Johnny Scott [313].

3rd d. Snowball [314] by Victor [122].

4th d. Blooming Heather 2nd [80] by Dred [15].

5th d. *Blooming Heather* [53] by *Moss Trooper 2nd* [48] (151).

6th d. *Mary* by *Fergy* (19).

7th d. *Jane* by *Bob.*

[624] *DAISY SECOND OF TARBREOCH.*

Calved January 2nd, 1880.

Bred by Mr. James Cunningham, Tarbreoch, Dalbeattie, Scotland; imported by and the property of Mr. Geo. Whitfield, Rougemont, Que.

Sire *Knowsley* (1279).

Dam *Daisy of Aireyolland* (2186), bred by Mr. Alex. McWhinnie, Aireyolland, Port William, Scotland.

[625] DAISY THIRD OF TARBREOCH.

Calved January 5th, 1881.

Bred by and the property of Mr. George Whitfield, Rougemont, Quebec.
Sire *Knowsley* (1279).
Dam *Daisy of Aireyolland* (2186) bred by Mr. Alex. McWhinnie.

[626] DOROTHEA.

Calved May 20th, 1880.

Bred by and the property of Mr. George Whitfield, Rougemont, Quebec.
Sire *Knowsley* (1279).
Dam *Meg Merrilies of Knocklae* [622] (2966) by *Neil Gow* (1138).
2nd d. *Lily of Knocklae* (2867) by *Blucher* (1136).
3rd d. *Old Lily*.

[627] *ELLA OF CHAPELHILL* (3369).

Calved March 14th, 1879.

Bred by Mr. Maxwell Clark, Chapelhill, Hawick, Scotland; imported by and the property of Mr. George Whitfield, Rougemont, Quebec.
Sire *Queensberry* (1027).
Dam *Baroness of Culmain* (2780) by *Mangerton* (525).
2nd d. *Miss Crocket*.

[628] *GEORGINA SECOND OF TARBREOCH.*

Calved March 10th, 1876.

Bred by Mr. Jas. Cunningham, Tarbreoch, Dalbeattie, Scotland; imported by and the property of Mr. Geo. Whitfield, Rougemont, Quebec.
Sire *Chamberlain* (666).
Dam *Georgina of Tarbreoch* (2190) from the herd of Mr. John Graham, of Shaw, Lockerbie, Scotland.

[629] JEAN.

Calved February 12th, 1881.

Bred by and the property of Mr. Geo. Whitfield, Rougemont, Quebec.
Sire *Laddie* [626].
Dam *Georgina 2nd of Tarbreoch* [628] by *Chamberlain* (666).
2nd d. *Georgina of Tarbreoch* (2190) from the herd of Mr. Graham.

COWS.

[630]   JEANNIE DEANS 3RD OF TARBREOCH.

Calved January 21st, 1880.

Bred by and the property of Mr. Geo. Whitfield, Rougemont, Quebec.

Sire *Prince Victor* (1473).
Dam *Jeannie Deans 2nd of Tarbreoch* (3599) by *Lord of Nithsdale* (616).
2nd d. *Jeannie Deans* (2696) by *Tom*.

[631]   *LADY QUEEN SECOND.*

Calved February 6th, 1880.

Bred by Mr. James Cunningham, Tarbreoch, Dalbeattie, Scotland; imported by and the property of Mr. George Whitfield, Rougemont, Quebec.

Sire *Knowsley* (1279).
Dam *Lady Queen* (3277) by *Sir James of Laws Hall* (826).
2nd d. *Mary of Laws Hall*, from the herd of Peter Morton, Loughbourn, Scotland.

[632]   *MEG MERRILIES OF KNOCKLAE* (2966).

Calved April 5th, 1876.

Bred by Mr. Thos. Currie, Knocklae, Balmaclellan, Scotland; imported by and the property of Mr. Geo. Whitfield, Rougemont, Quebec.

Sire *Neil Gow* (1138).
Dam *Lily of Knocklae* (2867) by *Blucher* (1136).
2nd d. *Old Lily*.

[633]   MISS NEILSON.

Calved June 1st, 1880.

Bred by and the property of Mr. Geo. Whitfield, Rougemont, Quebec.

Sire *Knowsley* (1279).
Dam *Georgina 2nd of Tarbreoch* [628] by *Chamberlain* (666).
2nd d. *Georgina of Tarbreoch* (2190) from the herd of Mr. Graham.

[634]   *VIOLET OF CHAPELHILL* (3671).

Calved February 12th, 1879.

Bred by Mr. Maxwell Clark, Culmain, Crockerford, Scotland; imported by and the property of Mr. George Whitfield, Rougemont, Quebec.

Sire *Queensberry* (1027).
Dam *Lizzie of Culmain* (2778) by *Clifton*.
2nd d. *Cherry*, from the herd of Mr. John Cannon, Congeith, Kirkgunzeon, Dumfries, Scotland.

[635]   BELL MAHONE.

Calved March 17th, 1881.

Bred by and the property of Mr. J. J. Bush, Lansing, Mich., U. S.
Sire Blackbird [636].
Dam Belle of Ingham [636] by Robin Hood [349].
2nd d. Lady Black [648] by Robin Hood [349].
3rd d. Jettie [348] by Our John [106].
4th d. Jet [343] by Prince Albert [190].
5th d. *Black Bess* [6] Imported.

[636]   BELLE OF INGHAM.

Calved May 6th, 1877.

Bred by Mr. J. N. Smith, Bath, Co Clinton, Mich., U. S.; the property of Mr. J. J. Bush, Lansing, Mich., U. S.
Sire Robin Hood [349].
Dam Lady Black [648] by Robin Hood [349].
2nd d. Jettie [348] by Our John [106].
3rd d. Jet [343] by Prince Albert [190].
4th d. *Black Bess* [6] Imported.

[637]   INGHAM BEAUTY.

Calved May 3rd, 1876.

Bred by Mr. J. N. Smith, Bath, Co. Clinton, Mich, U. S., the property of Mr. J. J. Bush, Lansing, Mich., U. S.
Sire Robin Hood [349].
Dam Maggie Lauder [110] by Black Jock [18].
2nd d. Empress [50] by *Jock* [10].
3rd d. *Beauty* [11] Imported.

Produce: 1878, Lady Mitchell [649]; 1879, Madeline [639]; 1880, Laura [650]; 1881, Mabel [638].

[638] MABEL.
Calved March 24th, 1881.
Bred by and the property of Mr. J. J. Bush, Lansing, Mich., U.S.
Sire Blackbird [633].
Dam Ingham Beauty [637] by Robin Hood [349].
2nd d. Maggie Lauder [110] by Black Jock [18].
3rd d. Empress [50] by *Jock* [10].
4th d. *Beauty* [11] Imported.

[639] MADELINE.
Calved April 7th, 1879.
Bred by and the property of Mr. J. J. Bush, Lansing, Mich., U.S.
Sire Blackbird [633].
Dam Ingham Beauty [637] by Robin Hood [349].
2nd d. Maggie Lauder [110] by Black Jock [18].
3rd. d. Empress [50] by *Jock* [10].
4th d. *Beauty* [11] Imported.

Produce: 1881, Pet Bush [642].

[640] MINNIE.
Calved June 14th, 1880.
Bred by Mr. George Coleman, Howell, Co. Livingston, Mich., U.S.; the property of Mr. J. J. Bush, Lansing, Mich., U.S.
Sire Rob Roy [501].
Dam Maid of Livingston [616] by Napier [346].
2nd d. Pride of the Dairy [497] by Hardfortune 2nd [255].
3rd d. Margaret [285] by *Dryfe* [222].
4th d. Black Bess [38] by Black Tom [37].
5th d. *Bell* [9] Imported.

Produce: 1882, Montrose [630].

[641] MONAH.
Calved May 4th, 1879.
Bred by Mr. Geo. Coleman, Howell, Co. Livingston, Mich., U.S.; the property of Mr. J. J. Bush, Lansing, Mich., U.S.
Sire Rob Roy [501].
Dam Chloe [584] by Herd Laddie [494].
2nd d. Pride of the Dairy [497] by Hardfortune 2nd [255].
3rd d. Margaret [285] by *Dryfe* [222].
4th d. Black Bess [38] by Black Tom [37].
5th d. *Bell* [9] Imported.

Produce: 1882, Burns [628].

[642] PET BUSH.

Calved February 10th, 1881.

Bred by and the property of Mr. J. J. Bush, Lansing, Mich., U.S.

Sire Blackbird [633].
Dam Madeline [639] by Blackbird [633].
2nd d. Ingham Beauty [637] by Robin Hood [349].
3rd d. Maggie Lauder [110] by Black Jock [18].
4th d. Empress [50] by *Jock* [10].
5th d. *Beauty* [11] Imported.

[643] BONNIE BELLE.

Calved August 11th, 1882.

Bred by Mr. Philo Lasher, Coffeysburg, Co. Davies, Mo., U.S.; the property of Messrs. A. M. & A. W. Callaham, Topeka, Kansas, U.S.

Sire Johnny Cope 2nd [456].
Dam Lady Belle [450] by Johnny Cope [265].
2nd d. Blue Bell [438] by Duncan [437].
3rd d. Belle Mahone [196] by Our John [106].
4th d. Nelly Grey [109] by Black Jock [18].
5th d. *Maggie* [48] by *Moss Trooper 2nd* [48] (151).

[644] LADY PRINCESS.

Calved March 7th, 1878.

Bred by Mr. Wm. Edie, Dunnville, Ont.; the property of Messrs. L. S. & W. B. Hall, Wakeman, Co. Huron, Ohio, U.S.

Sire Nelson [435].
Dam Gipsy Queen [194] by Our John [106].
2nd d. Dido [88] by Victor [122].
3rd d. Magnolia [84] by Black Jock [18].
4th d. Galloway Lass [42] by *Jock* [10].
5th d. *Jet* [2] Imported.

Produce: 1881, Duke of Argyle [681]; 1882, Huron Lad [632].

[645]   CREOLE.

Calved May 26th, 1881.

Bred by Mr J. J. Bush, Lansing, Mich., U. S.; the property of Messrs. Parker & Hardy, Abingdon, Ill., U. S.

Sire Rob Roy [501].
Dam Sylph [646] by Robin Hood [349].
2nd d. Lady Black [648] by Robin Hood [349].
3rd d. Jettie [348] by Our John [106].
4th d. Jet [343] by Prince Albert [190].
5th d. *Black Bess* [6] Imported.

[646]   SYLPH.

Calved June 9th, 1878.

Bred by Mr. J. J. Bush, Lansing, Mich., U. S.; the property of Messrs. Parker & Hardy, Abingdon, Ill., U. S.

Sire Robin Hood [349].
Dam Lady Black [648] by Robin Hood [349].
2nd d. Jettie [348] by Our John [106].
3rd d. Jet [343] by Prince Albert [190].
4th d. *Black Bess* [6] Imported.

[647]   CAPITOLA.

Calved December 30th, 1882.

Bred by Mr. Geo. Coleman, Marion, Co. Livingston, Mich., U.S; the property of Mr. S. A. Browne, Pentwater, Mich., U.S.

Sire Blackbird [633].
Dam Marion [614] by Rob Roy [601].
2nd d. Pride of the Dairy [497] by Hardfortune 2nd [255].
3rd d. Margaret [285] by *Dryfe* [222].
4th d. Black Bess [38] Black Tom [37].
5th d. *Bell* [9] Imported.

[648]   LADY BLACK.

Calved April 10th, 1874.

Bred by Mr. J. N. Smith, Bath, Co. Clinton, Mich., U.S.; the property of Mr. S. A. Browne, Pentwater, Co. Oceana, Mich., U. S.

Sire Robin Hood [349].
Dam Jettie [348] by Our John [106].
2nd d. Jet [343] by Prince Albert [190].
3rd d. *Black Bess* [6] Imported.

Produce: 1877, Belle of Ingham [636]; 1878, Sylph [646]; 1879, Princess [653]; 1880, Miss Black [652]; 1881, Darkness [657]; 1883, Blackwell [643].

[649] LADY MITCHELL.

Calved March 18th, 1878.

Bred by Mr. J. J. Bush, Lansing, Mich., U. S.; the property of Mr. S. A. Browne, Pentwater, Co. Oceana, Mich., U. S.

Sire Robin Hood [349].
Dam Ingham Beauty [637] by Robin Hood [349].
2nd d. Maggie Lauder [110] by Black Jock [18].
3rd d. Empress [50] by *Jock* [10].
4th d. *Beauty* [11] Imported.

Produce: 1881, Iona [629].

[650] LAURA.

Calved April 6th, 1880.

Bred by Mr. J. J. Bush, Lansing, Mich., U. S.; the property of Mr. S. A. Browne, Pentwater, Co. Oceana, Mich., U. S.

Sire Rob Roy [501].
Dam Ingham Beauty [637] by Robin Hood [349].
2nd d. Maggie Lauder [110] by Black Jock [18].
3rd d. Empress [50] by *Jock* [10].
4th d. *Beauty* [11] Imported.

Produce: 1882, Midnight [651].

[651] MIDNIGHT.

Calved November 19th, 1882.

Bred by and the property of Mr. S. A. Browne, Pentwater, Co. Oceana, Mich., U. S.

Sire Blackbird [633].
Dam Laura [650] by Rob Roy [501].
2nd d. Ingham Beauty [637] by Robin Hood [349].
3rd d. Maggie Lauder [110] by Black Jock [18].
4th d. Empress [50] by *Jock* [10].
5th d. *Beauty* [11] Imported.

[652] MISS BLACK.

Calved August 1st, 1880.

Bred by Mr. J. J. Bush, Lansing, Mich., U. S.; the property of Mr. S. A. Browne, Pentwater, Co. Oceana, Mich., U.S.

Sire Blackbird [633].
Dam Lady Black [648] by Robin Hood [349].
2nd d. Jettie [348] by Our John [106].
3rd d. Jet [343] by Prince Albert [190].
4th d. *Black Bess* [6] Imported.

[653] PRINCESS.

Calved October 10th, 1879.

Bred by Mr. J. J. Bush, Lansing, Mich., U. S.; the property of Mr. S. A. Browne, Pentwater, Co. Oceana, Mich., U.S.

Sire Blackbird [633].
Dam Lady Black [648] by Robin Hood [349].
2nd d. Jettie [348] by Our John [106].
3rd d. Jet [343] by Prince Albert [190.]
4th d. *Black Bess* [6] Imported.

Produce: 1883, Anniversary [659].

[654] TOPSY 2ND.

Calved November 8th, 1882.

Bred by and the property of Mr. R. B. Caruss, St. Johns, Co. Clinton, Mich., U. S.

Sire Lord Lochinvar [405].
Dam Topsy [369] by Johnny Hood [354].
2nd d. Rosa McNeil [345] by Hardfortune [154].
3rd d. Lizzie [144] by Dred [15].
4th d. Bonny [74] by *Jock* [10].
5th d. *Chloe* [4] Imported.

[655] HELEN.

Calved June 1st, 1881.

Bred by Mr. Thos. McCrae, Guelph, Co. Wellington, Ont.; the property of Messrs. J. W. Duncan & Sons, Independence, Mo., U.S.

Sire Lord Lochinvar [405].
Dam Galloway Lass [234] by *Pride of the Speed* [159].
2nd d. Maümee [92] by *Jock* [10].
3rd d. *Newbie Lass* [75] Imported.

[656] MEG.

Calved September 10th, 1881.

Bred by Mr. Wm. P. Darrow, Jefferson, Co. Hillsdale, Mich., U.S.; the property of Mr. H. A. Disbrow, Atlantic, Co. Cass, Iowa, U.S.

Sire Rob Roy [501].
Dam The Nymph [503] by Shoo-Fly [184].
2nd d. Queen of the Dairy [495] by Our John [106].
3rd d. Maggie Lauder [110] by Black Jock [18].
4th. d. Empress [50] by *Jock* [10].
5th d. *Beauty* [11] Imported.

[657] DARKNESS.

Calved August 9th, 1881.

Bred by Mr. Samuel A. Browne, Pentwater, Mich., U.S.; the property of Messrs. David Corey & Son, Yates City, Ill., U.S.

Sire Blackbird [633].
Dam Lady Black [648] by Robin Hood [349].
2nd d. Jettie [348] by Our John [106].
3rd d. Jet [343] by Prince Albert [190].
4th d. *Black Bess* [6] Imported.

[658] NELL OF WEA.

Calved March 1st, 1883.

Bred by and the property of Messrs. Isaac B. Lutz & Son, Lafayette, Co. Tippecanoe, Ind., U.S.

Sire Bob Wooley [410].
Dam Dora 2nd [293] by Shoo-Fly [184].
2nd d. Dora [288] by Shoo-Fly [184].
3rd d. Dairymaid [286] by *Dryfe* [222].
4th d. Black Bess [38] by Black Tom [37].
5th d. *Bell* [9] Imported.

[659] ANNIVERSARY.

Calved March 5th, 1883.

Bred by and the property of Mr. S. A. Browne, Pentwater, Mich., U. S.

Sire Blackbird [633].
Dam Princess [653] by Blackbird [633].
2nd d. Lady Black [648] by Robin Hood [349].
3rd. d. Jettie [348] by Our John [106].
4th d. Jet [343] by Prince Albert [190].
5th d. *Black Bess* [6] Imported.

[660] ALICE MAY.

Calved February 29th, 1883.

Bred by and the property of Mr. Peter Davy, Monterey, Co. Waukesha, Wis., U. S.

Sire *Bothwell* [431].
Dam Isabel [282] by Roger [198].
2nd d. Sis [203] by Hardfortune [154].
3rd d. Susan [157] by *Dryfe* [222].
4th d. Bess [125] by Marquis [19].
5th d. *Chloe* [4] Imported.

INDEX.

BREEDERS AND OWNERS.

Anderson, George, Stanley, 4, 73.
Armitage, Thomas, Colchester, 5.
Armstrong, Wm., Foster Meadow, Scotland, 158.
Ashbaugh, Z. R., Hillsdale, Mich., U.S., 156.

Beattie, James, Newbie House, Annan, Scotland, 7, 17, 71, 72, 75.
Beattie, Joseph, Newbie House, Annan, Scotland, 4.
Beattie, Wm., Newbie House, Annan, Scotland, 7.
Beattie, Simon, Annan, Scotland, 55, 56, 72, 165, 166, 167, 168.
Biggar & Son, Thomas, Chapelton, Dalbeattie, Scotland, 46, 141, 142, 152, 153, 159, 162.
Borer, John, West Flamboro', 100, 101.
Boulton, Henry J., Humberford, 73.
Bovell, Dr., Toronto, 98.
Broach, J., Waterbeck, Dumfriesshire, Scotland, 165, 166.
Brown, Robert, Capaylann, Dumfries, Scotland, 65.
Browne, Samuel A., Pentwater, Mich., U.S., 62, 169, 177, 178, 179, 180.
Burkholder, Isaac, Waterloo, 95, 104, 107.
Burleigh, G. S., Mechanicsville, Iowa, U.S., 141, 142.
Bush, John J., Lansing, Mich., U.S., 60, 61, 168, 174, 175, 176, 177, 178, 179.

Callaham, A. M. & A. W., Topeka, Kansas, U.S., 176.
Calvert, James, Dromore, 50, 151.
Cannon, John, Dumfries, Scotland, 56, 161.
Carruthers, Francis, Cleughheads, Lockerbie, Scotland, 154, 155, 165.
Carruthers, John, Belcathill, Lockerbie, Scotland, 162.
Carruthers, M., Kirkhill, Dumfries, Scotland, 1, 2, 13, 65, 66, 67, 154.
Carter, William, San Belle Plains, Kansas, Mo., U.S., 64.
Caruss, R. B., St. John, Mich., U.S., 33, 34, 38, 39, 49, 54, 120, 121, 122, 124, 125, 126, 149, 150, 159, 162, 179.
Chaffey, Thomas, Acton, 70.
Chambers, J. P., Lousay House, Abbey Town, Carlisle, 51.
Charlton, James, Lobo, 17.
Charlton, Joseph, Duncrief, 18, 22, 90, 91, 104.
Clark, Maxwell, Chapelhill, Scotland, 172, 174.
Clelland, William, Martinville, Mo., U.S., 45.
Clinkscales, E. C., Columbia, Mo., U.S., 487.
Coleman, Floyd, St. John, Co. Clinton, Mich., U.S., 149.
Coleman, George, Howell, Mich., U.S., 49, 54, 56, 57, 60, 61, 62, 63, 146, 163, 164, 168, 169, 170, 175, 177.
Coleman, Jasper, St. John, Co. Clinton, Mich., U.S., 146, 147.
Coleman, John, West Flamboro', 15, 88, 102.
Collacutt, Mr., Darlington, 73.
Conway, Robert, Co. York, 74, 75, 76.
Corey & Son, David, Yates, Co. Knox, Ill., U.S., 156, 180.

Cotter, George, Nelson, 4.
Cunningham, James, Tarbreoch, Dalbeattie, Scotland, 46, 55, 154, 160, 163, 165, 171, 172, 173.
Cunninghame, R. DeBarre, Honsol, Kirkcudbright, Scotland, 56, 166.
Currie, Thomas, Knocklae, Scotland, 173.

Darrow, William P., Jefferson, Mich., U.S., 54, 146, 147, 148, 156, 163, 179.
Davy, Peter, Ashippun, Wis., U.S., 24, 25, 27, 28, 29, 32, 35, 36, 41, 42, 51, 55, 57, 58, 97, 108, 111, 112, 113, 116, 117, 119, 120, 126, 127, 148, 152, 153, 154, 155, 156, 165, 171, 180.
Denison, Col. R. L., Toronto, 10, 20, 21, 81, 99, 100, 101.
Devlin, Henry, Guelph, 130, 131.
DeWitt, Anderson, Washington Court House, Ohio, U.S., 39, 64, 110.
Disbrow, H. A., Atlantic, Iowa, U.S., 49, 63, 146, 147, 179.
Dow, William, Nichol, 86, 104, 135.
Duke of Buccleuch, K.G., Scotland, 45, 53.
Duncan & Sons, J. W., Independence, Mo., U.S., 63, 179.
Dunlop, John, East Zorra, 7, 8, 68, 69, 82.

Earl of Galloway, Garlieston, Scotland, 54, 60, 142, 157, 160, 163, 168.
Earl of Selkirk, St. Mary's Isle, Kirkcudbrightshire, Scotland, 6, 87.
Edie, William, Dunnville, 44, 48, 136, 137, 140, 141, 176.
Elliott, Andrew, Galt, Co. Waterloo, 48, 144.

Ferguson, John, Kilquhanity, Kirkpatrick, Scotland, 161.
Fisher, Thomas, Craignarget, Wigtonshire, Scotland, 157, 160, 168.
Fleming, John, Vaughan, 1, 3, 21, 66, 69, 70.
Fraser, Mrs., Glaisters, Kirkpatrick, Scotland, 158.

Giddings, L. Romeyn, Downers Grove, Ill., U.S., 54.
Giles, J., Boston, Mass., U.S., 16, 87.
Gordon, Sir William, Earlston, Scotland, 47, 56.
Graham, George, Riggfoot, 7.
Graham, James, Vaughan, 2, 15, 21, 67, 68, 78, 83.
Graham, John, Shaw, Lockerbie, Scotland, 90, 91.
Graham, John, Kirkconnel, Castle Douglas, Scotland, 46, 141, 143.
Graham, Wm., Lockerbie, Scotland, 16.
Graham, Wm., Shaw, Dryfe, Dumfries, Scotland, 21, 65, 67.
Graham, Wm., Vaughan, 1, 2, 49, 65, 66, 67, 68, 70, 71, 75, 83, 91, 100.
Graham, W. R., Vaughan, 2, 3, 4, 7.
Greig, George, Dalry, Galloway, Scotland, 55.
Grierson, James, Caigton Hall, Dalbeattie, Scotland, 3.
Grieve, Gilbert, Minnydow, Dalbeattie, Scotland, 55.
Griffith, William, Little Current, Manitoulin, 64.

Hall, L. S. & W. B., Wakeman, Ohio, U.S., 61, 62, 176.
Hamilton, Ashly, Butler, Mo., U.S., 58, 59, 171.
Harrison, A., Mangerville, N.B., 20, 98, 99.
Hart, R. G., Lapeer, Mich., U.S., 18, 26, 28, 30, 58, 64, 108, 109, 110, 113, 114, 119.
Hickson, Joseph, Montreal, Que., 30, 50, 107, 112, 118, 150, 151.
Higginbotham, Col. N., Guelph, 53.
Holden, Edmund, Grape Island, W. Va., U.S., 61.
Hood, George, Guelph, 30, 129.
Hood William, Guelph, 10, 14, 17, 18, 21, 22, 23, 24, 26, 33, 38, 42, 47, 63, 81, 82, 90, 94, 95, 96, 101, 102, 103, 104, 105, 106, 108, 111, 120, 121, 128, 129, 138, 144, 146, 147.
Hurren, William, Mountsberg, 17.
Hurlbert, J. E., Eldora, Iowa, U.S., 36.

INDEX.

Jardine, Joseph, Saltfleet, 13, 18, 66, 84.
Juliand, Joseph, Bainbridge, N.Y., U.S., 118.

Keith, William, Pittsford, Mich., U.S., 54, 164.
Kerr, Alexander, Westminster, 68, 90, 91.

Laidlaw, James, Guelph, 88.
Laidlaw, John, Guelph, 102.
Lasher, Philo, Coffeysburg, Mo., U.S., 44, 45, 48, 57, 58, 136, 137, 138, 139, 170, 176.
Lutz & Son, Isaac B., Lafayette, Ind., U.S., 64, 180.
Lyons, Jeremiah, Dundas, 18.

Martin, Samuel, Eldora, Iowa, U.S., 57, 148.
Matthews, A. B., Kansas City, Mo., U.S., 36, 37, 38, 39, 127, 128, 129, 130, 131, 139.
Maxwell, Wm., Castle Douglas, Scotland, 154, 155.
Messenger, David, Cooksville, 74, 82.
Miller, George, Markham, 66, 79, 80, 85, 92, 99.
Miller, L., Marysville, Mo., U.S., 48.
Miller, William, Pickering, 71.
Millican, John, Wedholme House, Abbey Town, Carlisle, 51, 153.
Milligan, Andrew, Corwall, Port William, Scotland, 52, 163.
Mitchelson, H. H. & J. Lynes, Brompton, England, 51.
Moffatt, Samuel, Aird, Crossmichael, Scotland, 52.
Monck, Wesley W., Martinville, Mo., U.S., 45.
Montgomery, W. B., Starkville, Miss., U.S., 30, 118, 119.
Mounsey, Alexander, Etobicoke, 12, 13, 83, 84.
Muir, John, Highbae, Castle Douglas, Scotland, 9.
Murphy, Michael, Enniskillen, 76.
McCrae, Thomas, Guelph, 1, 9, 12, 13, 14, 15, 16, 17, 18, 19, 22, 23, 24, 25, 27, 29, 30, 31, 33, 36, 37, 38, 43, 45, 46, 49, 50, 52, 53, 54, 63, 64, 67, 68, 71, 74, 75, 79, 80, 83, 85, 86, 87, 88, 90, 91, 92, 93, 94, 97, 98, 103, 106, 107, 127, 128, 129, 131, 132, 133, 134, 141, 142, 143, 145, 157, 158, 159, 160, 161, 162, 163, 179.
McCrae, William, Manor Bank, Guelph, 23, 39, 46, 50, 128, 130, 131, 150, 151.
McCulloch, John, Port Elgin, 15, 89.
McDaniels, J., Bath, Mich., U.S., 34, 126.
McFadyean, Robert, New Galloway, Scotland, 166.
McHardy, F., Emporia, Kansas, U.S., 45, 46, 47, 48, 64, 141, 142, 143, 144, 145.
McKay & Burleigh, Mechanicsville, Iowa, U.S., 111, 136, 137.
McLean, James, Annan, Scotland, 167, 168.
McLean, John, Clover Hill, 3.
McLeod, Norman, King, 68.
McMicken, William, Gatehouse of Fleet, Scotland, 157, 158.
McNeil, Arthur, Vaughan, 3, 9, 11, 15, 16, 19, 20, 22, 32, 42, 70, 82, 88, 89, 97, 104, 108, 109, 121.
McPherson, D., Puslinch, 89.

Neilson, Joseph, Killimingan, Dumfries, Scotland, 160.
Nivison, J. S. & A., Lairdlaugh, Dalbeattie, Scotland, 60.

Palmer & Sons, Boscobel, Wis., U.S., 48, 132.
Parker & Hardy, Abingdon, Ill., U.S., 177.
Paterson, J. Jardine, Balgray, Scotland, 161, 162.
Percy, Hugh, Josselyn, Eskrigg House, Scotland, 155.
Peterson, H. W., Hawkesville, 14, 26, 79, 94, 95, 102.
Platt, M. R., Kansas City, Mo., U.S., 39, 40, 41, 42, 43, 44, 106, 121, 123, 124, 125, 132, 133, 134, 135.

Quarry, A., Guelph, 14.

Rae, Robert, Meikle, Cocklick, Dalbeattie, Scotland, 159.
Robson, George, Markham, 2.
Roddick, George, Cobourg, 2, 6, 8, 75, 87.
Roddick, William, Cobourg, 10, 82.
Rogers, J. J., Abingdon, Ill., U.S., 59, 171.
Rogerson, William, Leighton Hall, Dumfries, Scotland, 1, 66.
Routledge, William, Port William, Wigtonshire, Scotland, 52, 53, 161.

Sanson, D., Olin, Co. Jones, Iowa, U.S., 44.
Saxby, Charles M., Freeport, Ill., U.S., 56.
Scott & Thrall, Co. Greenwood, Kansas, U.S., 39.
Shannon, William, Barmoffity, Dalbeattie, Scotland, 142, 159.
Shennan, W. & J., Balig, Kirkcudbright, Scotland, 7, 52, 75, 82.
Smart, Robert, Dalbeattie, Scotland, 53.
Smith, J. N., Bath, Mich., U.S., 28, 29, 32, 33, 34, 39, 43, 53, 62, 120, 121, 122, 123, 124, 125, 126, 136, 137, 145, 159, 162, 174, 177.
Snell, John, Chinguacousy, 1, 4, 5, 6, 8, 9, 10, 11, 12, 69, 71, 72, 73, 74.
Snell, late John, Edmonton, 76, 77, 78, 80, 81.
Sproat, William, Borness Borgue, Kirkcudbright, Scotland, 11.
Steel, William, Short Riggs, Scotland, 167.
Steinmetz, Daniel, Shoeneck, Pa., U.S., 43, 103, 136, 168.
Sutherland, J. B., Petite Cote, 28, 113, 114.

Tew, H. C., West Flamboro', 100, 101, 105.
The Kansas State Agricultural College, Manhattan, Kansas, U.S., 33, 137, 139.
The Nebraska State University, U.S., 41, 43, 137.
The Michigan State Agricultural College, Lansing, Mich., U.S., 12, 28, 33, 76, 114, 115, 116, 120.
Thompson, Col. E. W., Aikenshaw, 5, 81.
Thomson, Adam, Castle Douglas, Scotland, 165.
Torrence, John, Vaughan, 4, 5, 66, 71, 72, 73, 79, 98.
Tubbs, S. A., Delhi, Mich., U.S., 169.

Vassie, John, Ancaster, 15.

Walker, William, Saltfleet, 13.
Wallace, Robert, Kirkcudbright, Scotland, 166, 167.
Weiste, Jacob, Co. Lancaster, Pa., U.S., 40, 41, 42, 43, 133, 135.
Welding, Edward, Vaughan, 4.
Whitfield, George, Rougemont, Que., 59, 60, 171, 172, 173, 174.
Whittaker, Capt. Jack, Oconomewoc, Wis., U.S., 24.
Wilcox, Allan, Co. Peel, 7.
Wilson, Christopher, Fergus, 47, 144.
Wilson, John, Westminster, 13, 84.
Wilson, M. & J. S., New Galloway, Scotland, 157, 159.
Wood, Charles C., Bureau, Ill., U.S., 63, 170.

INDEX TO BULLS.

	NO.	PAGE.
Aberdeen	[427]	40
Aikendrum	[133]	13
Allan	[623]	59
Arabi Bey	[548]	52
Argyle	[176]	18
Autumn (1698)	[549]	52
Badger	[325]	29
Badger Boy	[296]	27
Baron	[590]	55
Ben	[381]	96
Bertie	[163]	17
Bill	[375]	35
Billy Hood	[354]	33
Billy McNeil	[327]	29
Bismarck	[393]	37
Bismarck	[638]	63
Blackamoor	[391]	37
Blackbird	[633]	62
Black Dan	[491]	48
Black Dane	[641]	64
Black Diamond	[588]	54
Black Douglas	[148]	15
Black Jock	[17]	1
Black Jock	[449]	44
Black Jock 2nd	[253]	23
Black Prince	[152]	15
Black Prince	[160]	16
Black Tom	[37]	3
Black Tom	[279]	26
Black Wellington	[169]	18
Blackwood	[355]	33
Bob	[252]	22
Bob Wooley	[410]	39
Bonnie Dundee	[136]	13
Bothwell	[431]	41
Buffalo Bill	[289]	26
Burns	[628]	60
Captain	[308]	28
Cariboo	[90]	8
Centennial	[335]	30
Cetewayo	[392]	37
Chapelton	[550]	52
Charley Cope	[457]	45
Chief Templar (1782)	[529]	51
Chub	[115]	11
Clandeboye	[272]	25

	NO.	PAGE.
Clansman	[624]	59
Comet	[341]	32
Comet 2nd	[617]	57
Cromwell 2nd (1759)	[543]	51
Cully	[336]	31
Curly Bob	[44]	4
Curly Jock	[639]	64
Dainty Davy	[636]	63
Dalbeattie	[473]	46
Dalrymple	[331]	29
De Burre	[613]	56
Dick Peterson	[219]	20
Disraeli	[519]	59
Doctor Hornbook	[387]	36
Donald	[123]	12
Don Bernardo	[333]	30
Dover Court	[216]	20
Dred	[15]	1
Dryfe	[222]	21
Duckieston	[143]	14
Dufferin	[266]	25
Duke of Argyle	[631]	61
Duncan	[64]	5
Duncan	[437]	42
Earlston	[478]	47
El Dorado	[118]	11
El Hakim	[91]	8
Elrig	[551]	52
Emperor	[342]	32
Enchanter	[546]	51
Fred	[200]	19
Fred 2nd	[480]	47
Fred 4th	[481]	47
Fred Douglas	[46]	4
Galloway Boy	[426]	40
Galloway Tam	[70]	6
Galloway Tam (100)	[100]	9
Garibaldi	[130]	13
Garfield	[634]	62
Geordie	[108]	10
Geordie of Riggfoot (234)	[76]	7
Governor St. John	[413]	39
Grampian	[29]	3

INDEX.

Name	NO.	PAGE.
Hardfortune	[154]	15
Hardfortune 2nd	[255]	23
Heanan	[321]	29
Herd Laddie	[494]	49
Honest Tom 2nd	[238]	21
Hopewell	[441]	43
Huron Lad	[632]	62
Independence	[370]	34
Iona	[629]	61
Ivanhoe	[242]	22
Jacob Benner's Bull	[196]	19
Jim	[199]	19
Jim Crow	[430]	41
Jock	[10]	1
Jock 2nd	[45]	4
Jock 2nd	[207]	20
Jock-a-Dink	[254]	23
Jock Gray	[338]	31
Jock O'Bombie	[139]	14
John A	[102]	10
John Borland	[620]	58
John Brown	[429]	41
Johnny Cope	[165]	17
Johnny Cope	[265]	24
Johnny Cope 2nd	[283]	26
Johnny Cope 2nd	[456]	44
Johnny Cope 4th	[489]	48
Johnny Cope 5th	[618]	57
Johnny Hood	[354]	33
Johnny McNeil	[366]	34
Johnny Scott	[313]	28
Johnny Scott 3rd	[621]	58
King Charles	[625]	59
King Jacob	[428]	41
King Tom	[175]	18
King William	[490]	48
Lad	[589]	54
Laddie	[626]	60
Lafayette Prince	[642]	64
Laird of Burleigh	[516]	50
Lippincote	[220]	21
Little John	[168]	17
Lochiel	[591]	55
Lochinvar, Jr	[508]	49
Lofty of Waterside	[552]	53
Lord Byron	[23]	2
Lord Cecil	[257]	23
Lord Chelmsford	[521]	50
Lord Douglas	[406]	38
Lord Kenmure	[271]	25
Lord Lochinvar	[405]	38
Lord Malcolm	[614]	56
Lord Malcolm Second	[635]	63
Lord Monck	[218]	20
Lord Napier	[150]	15
Lord Starr	[334]	30
Lord Wellington	[332]	30
Lorne	[447]	43
Macgregor	[339]	31
Macleod of Dare	[471]	46
Macleod of Drumlanrig (1564)[470]		45
Macleod 2nd of Drumlonrig (1676)[553]		53
Major Gray	[273]	25
Malcolm	[62]	5
Maori Chief (1433) [592]		55
Marquis	[19]	2
Mochrum	[554]	53
Modock	[640]	64
Monitor	[121]	12
Montrose	[630]	61
Moreau	[622]	59
Morrach	[78]	8
Moss Trooper	[20]	2
Moss Trooper (296)	[74]	7
Moss Trooper	[135]	13
Moss Trooper 2nd (151)	[48]	4
Moss Trooper 4th	[493]	49
McNeil	[443]	43
McQuhoon	[26]	2
Napier	[346]	32
Nelson	[47]	4
Nelson	[435]	42
Nero 2nd	[371]	34
Nero 3rd	[373]	34
Norval	[72]	7
Our John	[106]	10
Peter	[116]	11
Peter	[593]	55
Peter Stewart	[390]	36
Premier of Lairdlaugh (1629)[727]		60
Pride of the Speed	[159]	16
Prince	[155]	16
Prince	[394]	37
Prince Albert	[190]	18
Prince Albert 2nd	[239]	21
Prince Charlie	[484]	48
Prince Lee Boo	[232]	21
Prince of Wales	[57]	5
Prince Oscar	[157]	15
Rob	[380]	35
Robbie Burns	[386]	96
Robert Bruce	[66]	6
Robin	[166]	17
Robin Hood	[144]	14
Robin Hood	[156]	16
Robin Hood	[349]	33
Robin Hood 2nd	[240]	22
Robin Hood 3rd	[256]	24

INDEX.

Name	NO.	PAGE.
Rob Roy	[298]	27
Roderick Dhu	[68]	6
Roderick Dhu	[637]	63
Roger	[198]	19
Rolly Polly	[616]	57
Royal Charlie	[474]	46
Saginaw	[94]	9
Saltfleet	[134]	13
Sam	[280]	26
Sambo	[71]	7
Sam of Garlieston (1610)	[555]	54
Scottish Chief	[300]	27
Scotia of Tarbreoch	[594]	56
Scotland	[21]	2
Selkirk	[101]	9
Shoo-Fly	[184]	18
Silver King	[439]	42
Sir Garnet	[267]	25
Sir John A	[137]	14
Sir Walter	[379]	35
Sir William	[340]	32
Sir William 2nd	[619]	58
Sir William Wallace	[138]	14
Snowball	[212]	28
St. Clair	[120]	11
Tam O'Shanter	[24]	2
Tecumseth	[117]	11
The Earl	[69]	6
The Earl	[472]	46
The Pilgrim (32)	[32]	3
The Warden	[263]	24
Tom	[395]	38
Tom, *alias* Tommy Dodd	[587]	54
Tom Scott	[305]	28
Tonawanda	[93]	9
Trooper	[530]	51
Uncle Joe	[425]	40
Uncle Sam	[440]	43
Uncle Tom	[58]	5
Uncle Tom	[107]	10
Uncle Tom	[113]	10
Victor	[122]	12
Victor Hugo	[92]	8
Waverley	[204]	19
Wellington 2nd	[404]	38
Wellington, *alias* Rob Roy	[162]	16
Wellington Boy	[264]	24
Wellington Boy	[432]	42
William Wallace, *alias* Clover Hill	[28]	3
Wonderful Lad	[214]	20
Young Blucher	[41]	3
Young Blucher	[164]	17
Young Dryfe	[127]	12
Young Franklin	[595]	56
Young Lochinvar	[303]	27
Young Matchless	[40]	3
Young Moss Trooper	[73]	7
Young Moss Trooper	[245]	22
Young Scotland	[63]	5
Young Solway	[411]	39
Zulu	[337]	31

INDEX TO COWS.

	NO.	PAGE.
Adah	[191]	95
Aggie	[359]	105
Agnes	[124]	83
Agnes 2nd	[116]	82
Alice Maud (3977)	[585]	154
Alice Maxwell	[596]	165
Alice May	[660]	180
Annie Laurie	[82]	77
Annie Laurie 2nd	[251]	104
Annie Sis	[262]	105
Anniversary	[659]	180
Barbara	[113]	92
Beauty	[11]	67
Beauty	[59]	73
Beauty	[151]	87
Beauty	[224]	101
Beauty	[258]	104
Beauty	[420]	133
Beauty	[597]	165
Beauty 2nd	[507]	148
Beauty of Guelph	[389]	129
Beauty Second of Troquhain (3517)	[556]	157
Bell	[9]	66
Bell	[536]	154
Bell	[598]	165
Bella	[149]	87
Belle of Ingham	[636]	174
Belle of Janefield	[388]	128
Belle Mahone	[196]	96
Bell Mahone	[635]	174
Bell of Cleughheads	[599]	165
Belted Lass	[557]	157
Bess	[125]	83
Bessie	[618]	170
Bessie Bell	[146]	86
Bessie Bell 2nd	[179]	92
Bessie Lee	[181]	93
Betty	[299]	112
Betty Laidlaw	[276]	107
Black Baby	[451]	138
Black Beauty	[500]	147
Black Bess	[6]	66
Black Bess	[38]	70
Black Bess	[193]	95
Black Bess	[260]	105
Black Bess	[459]	140

	NO.	PAGE.
Black Bess 2nd	[284]	108
Black Bessy	[453]	139
Blackie 2nd of Keels (4188)	[525]	152
Black Jane	[210]	99
Black Jess	[620]	170
Black Jessie	[218]	100
Black Maria	[212]	99
Black Nellie	[173]	91
Black Queen	[416]	132
Black Rachel	[172]	91
Black Rachel	[294]	110
Black Sall	[223]	101
Black Swan	[219]	100
Black Swan	[328]	118
Blacky	[8]	66
Blanche	[600]	165
Blankless	[330]	118
Bloom	[601]	166
Blooming Beauty	[54]	73
Blooming Heather	[53]	72
Blooming Heather	[336]	119
Blooming Heather 2nd	[80]	76
Blossom	[621]	171
Blossom of Tarbreoch	[558]	157
Blue Bell	[235]	102
Blue Bell	[438]	136
Bonnie Belle	[643]	176
Bonnie Bessie	[492]	145
Bonnie Lassie	[408]	131
Bonny	[1]	65
Bonny	[27]	69
Bonny	[136]	85
Bonny 2nd	[36]	70
Bonny 2nd	[74]	75
Bonny Lass	[295]	111
Border Belle	[522]	151
Border Lass	[384]	128
Brenda	[77]	76
Bright Eye	[424]	135
Capitola	[215]	99
Capitola	[647]	177
Careful	[559]	157
Carnation	[560]	158
Catharine	[339]	120
Cheerful	[561]	158
Cherry	[81]	76
Cherry	[307]	113

	NO.	PAGE.
Cherry Blossom	[602]	166
Cherry of Hensol	(3554) [603]	166
Chloe	[4]	65
Chloe	[584]	164
Clara	[126]	83
Claret Fourth	(4187) [526]	152
Clinton	[511]	149
Coila	[562]	158
Comfort	[563]	158
Cooderina	[161]	89
Coquette	[119]	83
Cora	[142]	85
Countess	[34]	70
Countess of Galloway	[468]	142
Craignarget	[611]	168
Creole	[645]	177
Cricket	[564]	159
Cricket	[209]	99
Curly	[436]	136
Curly Nell	[544]	156
Dagmar	[153]	88
Dairymaid	[51]	72
Dairymaid	[158]	89
Dairymaid	[286]	109
Daisy	[586]	164
Daisy Deane	[167]	90
Daisy Second of Tarbreoch	[624]	171
Daisy Third of Tarbreoch	[625]	172
Damson 2nd	[537]	154
Dandy	[69]	75
Dark Lady	[505]	148
Darkness	[657]	180
Dew Drop	[377]	127
Dido	[88]	78
Dinah	[613]	168
Dora	[288]	109
Dora 2nd	[293]	110
Dora Bell	[297]	111
Dora Deane	[615]	169
Dorothea	[626]	172
Duchess of Sutherland	[310]	113
Ebony	[177]	92
Edith of Troquhain	(3518) [565]	159
Ella of Chapelhill	(3369) [627]	172
Elsie	[128]	84
Elvira	[83]	77
Em	[144]	86
Empress	[50]	72
Empress	[270]	106
Empress	[517]	150
Eugenie	[184]	93
Eva	[93]	79
Fairy Dale	[482]	144
Fancy	[202]	97
Fanny	[129]	84
Fanny Blair	[477]	143
Flora	[22]	68
Flora	[141]	85
Florence	[374]	126
Galloway Lass	[42]	71
Galloway Lass	[234]	102
Georgina Second of Tarbreoch	[628]	172
Gipsy	[225]	101
Gipsy Queen	[89]	78
Gipsy Queen	[311]	114
Gipsy Queen 2nd	[194]	96
Grace	[99]	81
Grace Darling	[145]	86
Hannah	[131]	84
Hannah	[512]	149
Hannah 2nd	[513]	150
Heather Bell	[12]	67
Heather Bell 2nd	[25]	68
Heather Bell 3rd	[85]	77
Heather Bell 4th	[137]	85
Heather Bloom	[399]	130
Helen	[103]	81
Helen	[655]	179
Helen Douglas	[33]	69
Helen of Kirkconnel	[469]	143
Hyena	[185]	94
Idaho	[97]	80
Ingham Beauty	[637]	174
Isabel	[282]	108
Jane Peterson	[221]	101
Jane Seton 2nd	(3787) [467]	142
Jane Seton 3rd	[566]	159
Jean	[201]	97
Jean	[629]	172
Jeanette	[504]	148
Jeannie Deans	[243]	103
Jeannie Deans 2nd of Tarbreoch	(3599) [465]	141
Jeannie Deans 3rd of Tarbreoch	[630]	173
Jeannie	[152]	88
Jenny Lind	[32]	69
Jenny Lind	[448]	138
Jenny of Tarbreoch	[567]	159
Jessamine	[621]	171
Jessie	[49]	71
Jessie	[257]	104
Jessie 2nd	[320]	116
Jessie Miller	[98]	80
Jet	[2]	65
Jet	[343]	120
Jet	[398]	129
Jet 2nd	[403]	131
Jet 2nd	[352]	122
Jet 3rd	[358]	123
Jet 4th	[362]	124

INDEX.

Name	NO.	PAGE
Jettie	[348]	121
Jettie 2nd	[357]	123
Jettie 3rd	[361]	124
Juliand	[334]	118
Kate	[176]	91
Kate	[296]	111
Kate	[460]	140
Kate	[538]	154
Kenmure	[112]	82
Lady Bell	[278]	107
Lady Bell	[450]	138
Lady Black	[462]	141
Lady Black	[648]	177
Lady Dufferin	[275]	107
Lady Geills	[568]	160
Lady Gordon	[182]	93
Lady Gordon	[419]	133
Lady Gray	[423]	134
Lady Hamilton	[274]	106
Lady Hope	[612]	168
Lady Heron	[180]	92
Lady Isabella	[100]	80
Lady Jane	[446]	137
Lady Kenmure	[140]	85
Lady Kenmure	[604]	166
Lady Kerr	[528]	152
Lady Lisgar	[241]	103
Lady Marion	[186]	94
Lady Mary	[16]	67
Little Mary	[400]	130
Lady Maxwell	[414]	132
Lady Mitchell	[649]	178
Lady Princess	[644]	176
Lady Queen Second	[631]	173
Lady Shannon	[466]	142
Lassie	[445]	137
Laura	[650]	178
Lavinia	[55]	73
Lavinia	[329]	118
Libbie	[208]	98
Lily	[409]	131
Lily Dale	[164]	89
Lily Dale	[233]	102
Little Emily of Kirkconnel-	(3321) [463]	141
Lizzie	[114]	82
Louan	[87]	78
Louisa	[86]	78
Louise Lorne	[417]	132
Lucetta	(3441) [569]	160
Lucetta 2nd	[570]	160
Lucy	[31]	69
Lucy	[52]	72
Lucy	[324]	117
Lucy	[434]	135
Lucy	[502]	147
Mabel	[638]	175
Madeline	[639]	175
Maggie	[48]	71
Maggie	[64]	74
Maggie	[120]	83
Maggie	[452]	139
Maggie	[605]	167
Maggie 3rd	[364]	125
Maggie 4th	[356]	123
Maggie 4th	[509]	149
Maggie 5th	[360]	124
Maggie Laidlaw	[382]	127
Maggie Lauder	[110]	81
Maggie Lauder	[148]	87
Maggie Lauder 2nd	[347]	121
Maggie Lauder 3rd	[353]	122
Maggie 4th of Elrig	(3883) [572]	161
Maggie 2nd of Killimingan	(3877) [571]	160
Maggie Shaw	[476]	143
Magnolia	[84]	77
Magnolia 2nd	[397]	129
Magnolia 3rd	[401]	130
Maid of Dumfries	(4974) [573]	161
Maid of Livingston	[616]	169
Margaret	[285]	108
Margaret	[344]	121
Marion	[614]	169
Marjorie	[205]	98
Mary	[35]	70
Mary	[111]	82
Mary	[171]	91
Mary	[281]	108
Mary	[524]	151
Mary Grey	[65]	74
Mary Hay	[147]	87
Mary Hood	[383]	128
Mary Little	[421]	134
Matilda	[302]	112
Maud	[545]	156
Maude	[337]	119
Maumee	[92]	79
May Queen	[143]	86
May Queen	[292]	110
May Queen 2nd	[237]	103
Meg	[656]	179
Meg Hood	[479]	144
Meg of Congeith	[574]	161
Meg Merrilies of Knocklae	(2966) [632]	173
Midnight	[651]	178
Mina	[187]	94
Mink	[547]	156
Minnehaha	[156]	88
Minnie	[178]	92
Minnie	[640]	175
Miss Black	[652]	178
Miss Carruthers	[606]	167
Miss Hope 2nd of Balgray	[575]	161

	NO.	PAGE.
Miss Lauder	[488]	145
Miss Maude	[376]	127
Miss Neilson	[633]	173
Miss Nellie of Belcathill	[576]	162
Miss of Belcathill	[577]	162
Miss Steele	[523]	151
Miss Torrence	[204]	98
Mollie	[301]	112
Mollie	[461]	140
Mollie Darling	[291]	110
Molly	[17]	68
Molly	[583]	163
Monah	[641]	175
Moss Rose Fourth (3721)	[532]	153
Moss Rose 3rd of Wedholme (3562)	[531]	153
Mrs. Snell	[105]	81
Nancy	[76]	76
Nancy	[607]	167
Nancy Blair	[483]	144
Nancy Lee Second	[579]	162
Nancy Second of Balgray	[578]	162
Nancy of Tarbreoch	[580]	163
Naomi	[619]	170
Nellie	[160]	89
Nellie	[539]	155
Nellie	[608]	167
Nellie of Corwall (3885)	[581]	163
Nell of Wea	[658]	180
Nelly Grey	[109]	81
Nelly Grey	[269]	106
Nelly Grey 2nd	[396]	129
Nelly Grey 3rd	[402]	130
Nettie	[610]	168
Newbie Lass	[75]	75
Nigra	[499]	147
Nora	[372]	126
Ontario	[442]	136
Passion Flower	[422]	134
Pearl	[195]	96
Pet	[498]	146
Pet Bush	[642]	176
Phillis	[7]	66
Pocahontas	[60]	74
Pocahontas 2nd	[166]	90
Pocahontas 3rd	[487]	145
Polly	[514]	150
Polly Shaw	[174]	91
Prairie Flower	[415]	132
Prairie Lass	[261]	105
Pride	[322]	116
Pride	[520]	50
Pride of the Dairy	[497]	146
Primrose of Kirkconnel (3317)	[463]	141
Princess	[653]	179
Princess Mary	[540]	155

	NO.	PAGE.
Queen	[418]	133
Queen Bess	[407]	131
Queen Mab	[206]	98
Queen Mary	[486]	145
Queen of Beauty	[95]	79
Queen of Beauty 2nd	[197]	96
Queen of Eskrigg	[541]	155
Queen of the Blacks	[496]	146
Queen of the Dairy	[495]	146
Queen of the West	[43]	71
Queen of the West	[168]	90
Queen's Own of Ridgeway	[217]	100
Queen Vic	[236]	102
Ranee Third (4185)	[527]	152
Rosa	[326]	117
Rosa	[350]	122
Rosa 2nd	[355]	122
Rosa 3rd	[363]	124
Rosa McNeil	[345]	121
Rose	[132]	84
Rose	[306]	113
Rose	[385]	128
Rose 5th of Wedholme (3948)	[533]	153
Rosy	[510]	149
Rosy Hood	[188]	94
Rosy Hood 2nd	[359]	123
Rosy Hood 3rd	[367]	125
Rosy Hood 4th	[368]	125
Rowena	[244]	103
Ruby	[61]	74
Salina	[104]	81
Sall	[14]	67
Sall	[170]	90
Sall	[192]	95
Sall	[485]	144
Sall 2nd	[79]	76
Sally	[365]	125
Scotch Maid	[323]	117
Sea Bird	[534]	153
Shaw	[21]	68
Sis	[203]	97
Sleeping Maggie	[433]	135
Sloe Second	[542]	155
Snip	[317]	115
Snip 2nd	[338]	120
Snip 3rd	[623]	171
Snovir	[318]	115
Snow	[319]	116
Snowball	[314]	114
Snowbloom	[316]	115
Snowdrift	[444]	137
Snowflake	[315]	114
Susan	[30]	69
Susan	[157]	88
Susie	[515]	150
Sweet Afton	[585]	164
Sweet Alice	[617]	169

	NO.	PAGE.
Sylph	[646]	177
The Nymph	[503]	147
The Squaw	[518]	151
Topsy	[13]	67
Topsy	[200]	97
Topsy	[213]	99
Topsy	[287]	109
Topsy	[335]	119
Topsy	[369]	126
Topsy	[454]	139
Topsy	[506]	148
Topsy 2nd	[56]	73
Topsy 2nd	[654]	179
Topsy of Tarbreoch	[582]	163
Topsy Wopsy	[189]	95
Venus	[3]	65

	NO.	PAGE.
Venus	[73]	75
Victoria	[135]	84
Victoria	[150]	87
Victoria	[290]	109
Violet of Chapelhill	[3671] [634]	174
Wandering Nellie	[268]	106
Wandering Nellie	[277]	107
White Bag	[5]	66
White Bag	[220]	100
Wild Eye	[304]	112
Woolwich Queen	[96]	79
Woolwich Queen	[378]	127
Young Bonny	[39]	70
Young Cherry	[609]	167
Young Mary	[246]	104

www.ingramcontent.com/pod-product-compliance
Lightning Source LLC
Chambersburg PA
CBHW062213220526
45471CB00009B/3187